Islands and Cities in Medieval Myth, Literature, and History

Beihefte zur

MEDIAEVISTIK

Monographien, Editionen, Sammelbände

Herausgegeben von Peter Dinzelbacher

Band 14

PETER LANG

Frankfurt am Main · Berlin · Bern · Bruxelles · New York · Oxford · Wien

Andrea Grafetstätter
Sieglinde Hartmann
James Ogier
(eds.)

Islands and Cities in Medieval Myth, Literature, and History

Papers Delivered at the International Medieval Congress,
University of Leeds, in 2005, 2006, and 2007

PETER LANG
Internationaler Verlag der Wissenschaften

Bibliographic Information published by the Deutsche Nationalbibliothek
The Deutsche Nationalbibliothek lists this publication in the
Deutsche Nationalbibliografie; detailed bibliographic data is
available in the internet at http://dnb.d-nb.de.

Cover illustration:
The City of Jerusalem, in: Ebstorfer Weltkarte (detail),
mappa mundi, Germany, c. 1300;
Die Ebstorfer Weltkarte, Birgit Hahn Woehrle (ed.),
Kloster Ebstorf, 2nd ed. 1993.

ISSN 1617-657X
ISBN 978-3-631-61165-4

Contents

Editors' Preface

The studies presented in this book derive from a series of sessions held at the annual International Medieval Congress in Leeds, UK under the aegis of the Oswald-von-Wolkenstein Gesellschaft (Frankfurt am Main, Germany) and the Associazione di Cultura Medioevale (Trieste, Italy), all of them organised by Professor Sieglinde Hartmann (University of Würzburg, Germany). Four sessions, held from 2004 to 2006, bore the title "Islands of the World and the Seven Seas in Medieval Myth and History", and three in 2007 the title "Cities, Myths and Literatures". These sessions had many of the same participants and, as the reader will discover, resulted in many ways in contiguous lines of investigation.

The stated objective of the island sessions was the location of a "starting point for a new investigation into the possible impact that myths and other fictitious stories about insular wonderlands had on the reasons why medieval men and women undertook their various missions, searches and explorations that finally led to the discovery of the New World". Similarly, the cities sessions "intended to find new connections between ancient myths and medieval constructions of real or imagined cities in literature." From the start, the scope of the papers reached across the globe and spanned the entire Middle Ages as the topic drew in scholars not just from Western Europe, but from Eastern Europe, Asia, and America. The wide range of interests among the contributors becomes apparent already in this volume's table of contents.

Islands and cities have several important features in common: they consist of bordered places with a definable inside and outside; they provide a sense of commonality for the inhabitants; and they form a barrier against the Other. We see, for example, in Patrizia Mazzadi's analysis of Dante's island of Purgatory, that the island can represent a place of security while the souls await their ascent into heaven. At the same time, Odysseus's failed attempt to reach the island—while not knowing its nature—demonstrates the island's ability to reject the non-believer. Thus the island's protection parallels the function of the temples and shrines that surrounded the Japanese capital of Kyoto (cf. Yuko Tagaya's "Kyoto in Myth and Literature") by appeasing and thus repelling the unclean spirits of impurity and pollution.

Both islands and cities, as defensible areas, have a connection to sacrality, as becomes clear in several of this volume's contributions (James Ogier's "Myths, *Mujeres*, and Mexico"; both of Yuko Tagaya's studies; and Jacek Kowzan's

"Heavenly Jerusalem as a *locus amoenus* in Medieval and Early Modern Polish Literature"). Cultures across the world associated both islands and walled cities with notions of paradise and perfection in their literatures. In China, for example, paradise is found on Mt. P'eng-Lai. As the narrative was received in Japan, the location shifts over time from an island in the Eastern Sea of China to the moon (an island in the sky) to a sub-aquatic island (Yuko Tagaya's "Far Eastern Islands and Their Myths"). Jarosław Wenta ("Holy Islands and Their Christianisation in Medieval Prussia") demonstrates the staying power of notion of holiness during a transition from pagan religion to Christianity. In the Germanic world, islands also provide the backdrop for acts of treachery by turning the paradigm of protection on its head (James Ogier's "Islands and Skylands" and Sieglinde Hartmann's "Insular Myths in the *Nibelungenlied*"). Finally, several contributions limn visitor's reactions to and influence on medieval cities (Maria Dorninger's "The Island of Cyprus in Travel Literature of the Fourteenth Century"; Sieglinde Hartmann's "A Medieval Poet's Sense of Humour"; and Andrea Grafetstätter's "Foreign Culture in a Foreign Town").

The editors hope that the often unusual perspectives presented here will encourage others to grapple with the varieties of bounded space (whether by walls or by water) in world literatures and history. The commonalities are often more striking than the differences, reflecting a common human need for barriers against the outside world.

The editors wish to thank Dr Rebecca J. Davies for her proofreading and Professor Dinzelbacher for accepting this volume in his series "Beihefte zur MEDIAEVISTIK".

Andrea Grafetstätter, Sieglinde Hartmann, and James Ogier

Islands and Skylands: An Eddic Geography

While the narratives in the Old Icelandic Poetic Edda take place in a mythic environment, they nevertheless have an abundance of landscape detail, conjuring up rivers, mountains, and trees as scenery for the events unfolding in the verses.[1] Indeed, the first work in the collection, the *Völuspá*, provides a visually concrete world creation myth for the stories that follow, and landscape features play a role in shaping dramatic scenes such as the flyting match between Thor and Odin across a sound in the *Hárbarðsljóð*. Among these landscape details, however, there are few islands, neither *eyjar* nor *hólmar*. Because the landscape lacks terrestrial coordinates, the islands, even if they bear a name (*Hárbarðsljóð*'s Algrœn [Hl 16], *Helgakviða Hundingsbana I*'s Brandey and Heiðinsey [HHI 22]) stand in no clear relationship to their surroundings and function predominantly as places of warfare.

One island does stand out, however, because its very nature as an island plays an important role, both physically and metaphorically, in its poem: Sævarstaðr, the island on which the evil king Níðuðr holds captive the recently de-sinewed Völundr the smith in the Völundarkviða. Unlike the other islands in the Edda, Sævarstaðr is sufficiently contextualized in its poetic landscape to permit speculation about its general location in the mythic environment. That mythic environment has, as will be argued below, a basis in a culturally defined interpretation of the night-time sky and stands in a structured relationship with other, equally celestial Germanic mythologems. Placing Sævarstaðr in this framework sheds light on a few puzzling aspects of the Völund/Wēland complex.

Cultural Astronomy

The tools for locating the island rest on basic assumptions derived from cultural astronomy – the study of reflexes of native astronomies in the cultural record (e.g., Krupp; Aveni) – and embodied logic – the branch of semantics which describes the gestalt-images which underlie human reasoning and metaphoric capabilities (see below; Lakoff and Johnson 1980; Johnson 1987). Cultural astronomy makes the assumption that for at least the past million years, humans

---

[1] I wish to express my sincere thanks to Rory McTurk for saving me from several errors in this article. The remaining infelicities are, of course, my own.

have spent some portion of their nights looking up at the stars and attempting to find some order in the celestial patterns, and that the shapes that humans perceive in the apparent randomness of the heavens must necessarily provide the grist for basic cultural phenomena, among them myth, legend, and religion. In the words of Jouko Hautalan, "One can well understand that people of nature, of the tribe, of antiquity could not fail to pay great attention to the evening's and the night's overarching mighty sight, the eternal, unchanging starry heaven's brightly shimmering lights, and to take notice of the solemn phenomena's forever immutable, precisely repeating regularity, and furthermore, being human, to seek an explanation, a first cause for those phenomena" (cited in Pentimäinen 117 [my translation]).[2] As archaeologists identify more and more ancient observatories and astronomical tools (with the recently unearthed Nebra disk as a particularly spectacular example, cf. Meller), it is becoming increasingly clear that astronomical lore has played a role in human knowledge for millennia and that one should expect to find reflexes of it in every manifestation of human culture. Naturally, the lore varies from culture to culture according to latitude and the vagaries of temporal and geographic transmission; one must approach each culture in terms of its location on the earth and its cultural inheritance.

For Scandinavia, the problem of latitude arises immediately: the farther north one ventures, the less useful the summer months become for naked-eye astronomy, thus the Icelandic term *vetrarbraut* (winter road) for the Milky Way. Nevertheless, Scandinavia, and even the North Atlantic, has the occasional cloudless winter night, and, weather permitting, ancient Scandinavians had ample opportunity to observe the winter skies and their movements – not perhaps enough opportunity for a full-fledged astronomy in the manner of Greece, but enough for a sky-map and concomitant mythology. The long nights spent in the sheepfold or the tedium of a week-long Atlantic crossing surely inspired star-watching, especially in light of the agricultural and navigational benefits accruing from astronomical knowledge.

But naked-eye astronomy does not provide the whole picture; astronomical lore also derives from an older, inherited culture. The Scandinavians inherited from their Germanic and, indeed, Indo-European stock the astronomical lore the descendants of which appear in, e.g., Vedic and classical Mediterranean mythology. For all these reasons, Northern Europeans had access to a transmitted and culturally reinforced body of knowledge about the heavens, including a cast

---

[2] "Voi hyvin ymmärttää, että luonnonihminen, kansanihminen, muinaisajan ihminen ei ole voinut olla kiinnittämättä mita suurinta huomiota iltaisin ja öisin yllään kaareutuvaan mahtavaan näkyyn, ikuisen, muuttumattoman tähtitaivaan kirkkaina kimmeltäviin valoihin, ja panematta merkille sen juhlallisten ilmiöiden ajasta aikaan järkkymättömän täsmällisesti toistuvaa säännöllisyyttä, ja myös, koska hän on ollut ihminen, etsimätta selitystä, alkusyytä näille ilmiöille."

of planetary gods (Wotan/Oðinn/Mercury, Thor/Jupiter) with inherited — if sometimes redistributed – functions.

This last point – of cultural inheritance – raises old, vexed, and ultimately unanswerable questions about the mutation of cultural material in related oral societies. Scholars as far back as Ælfric [in the homily "De falsis diis"] have pondered, for example, the misalignments between the Germanic and the Graeco-Roman pantheons; if Odin corresponds to Mercury and Thor to Jupiter, then how can Odin have fathered Thor, when Jupiter fathered Mercury (proof, in Ælfric's mind, that the Germanic pagan religion was just so much hogwash)? But the similarities between the systems of divinities, rather than the differences, should matter more in comparative mythology, just as similarities rather than differences do in comparative linguistics. For language transmission, a culture needs a synchronically stable system of symbolic morphemes. For mythological transmission, it helps to have stable visible referents available on a regular basis. The Indo-European peoples, like other cultures the world over, made use of the oldest pedagogical materials and mnemonic devices known to humankind – the stars and planets – and anchored their mythologies in them. One may assume a transmission, from generation to generation, of Indo-European mythologems through the agency of a mythophor who used the skies as his text (or her text – the Edda begins with a *völva*, after all). In many cultures, this job falls to the shaman, who journeys into the heavens to gain enlightenment and must thus inculcate into his followers an understanding of his path, e.g., along the World Tree (Eliade 275 et passim). While it is prudent to avoid the vexed question of whether or not ancient Scandinavia had shamanism (cf. von Schnurbein; Dubois 53-54), the apparent identity of Germanic gods with planetary gods implies a role for astronomy in ritual life and consequently suggests that astral knowledge, be it shamanistic or otherwise, formed a part of priestly function.

The Northern Sky

Astronomically based ritual, however, tends to be syncretistic and need not assign one and only one meaning to each aspect of the heavens; in fact, the visual hook for mythologems is usually polyvalent. For example, the Milky Way can, according to context, represent 1) a stream of liquid, as in the story of Hera spraying milk across the sky, which gives us the term "Milky Way" in the first place; 1b) a path or way of some other material, as in Icelandic "vetrarbraut"; or 2) the World Tree, e.g., in its Scandinavian guise as Yggdrasill.[3] In this

---

[3] When speaking of Yggdrasill, it is best to avoid the early twentieth-century parlor game of determining whether the tree is an oak, an ash, a yew, a linden, or any other arborial species, and especially in the context of a culture which says proverbially

polyvalence, one can detect a confluence of traditions, some recognizably Indo-European and some not, but each with its own satellites of images. The galactic tree, for example, carries with it one or more birds in its crown, a hart or harts at its mid-section, and a serpent or serpents at its root (cf. Motz 1991, 135 for Yggdrasill; Schele 59ff. for the Maya World Tree). When the Milky Way becomes a path, the hart turns into a horse traveling upon the path, perhaps because a horizontal Milky Way lies under the constellation Pegasus, providing a road for the horse. Frequently, the images overlap and coalesce; the tree drips liquid from its branches (*Völuspá* 19), and the hart drips liquid from its horns (*Grímnismál* 26). The image need not be static; the Milky Way rotates in the sky, and the observer can change point of reference and direction of view; thus the celestial object can appear to rise up, tree-like, or lie flat as a path, just as it can arch across the center of the sky or hug the horizon, obviating the necessity of positing two different traditions for the World Tree as does Winterbourne (62-68). Furthermore, the interpretation of seemingly random clusters of stars depends on the observer's cultural knowledge. Europe and North America see a man in the moon; Latin America and China see a rabbit; the Chukchee see a girl with a bucket (Motz 1983, 380). One sees in the stars what one is told to see, and one is told to see what fits the story-teller's particular purpose when he or she uses the sky as a visual aid.

The two images, the stream and the tree, have a commonality which becomes very important in their interpretation: they are both directional. A stream flows in one direction, and a tree grows upward. Consequently, their very nature gives a framework of directionality to the rotating heavens, one independent of earth-bound coordinates. The crown of the tree, with its resident bird, must point "up," the roots, with their serpent, must point "down." One could argue that the part of the Milky Way that splits into two distinct "streams" at the Cygnus Rift represents the roots of the tree Yggdrasill; if one accepts this, it follows that the other end of the Milky Way visible in the Northern Hemisphere must represent the crown of the tree. The assumption of a separate and earth-independent grid for the revolving heavens also solves the difficulty of interpreting all the cardinal directions to which the Eddic poems make reference: "up" and "north" mean up the tree toward Gemini; "down" and "south" mean toward Scorpio and the underworld; "west" toward Pegasus; and "east" toward Ophiuchus. Since the "southern" constellations, which normally dominate the summer sky, fade into invisibility in the far north in the summer, their heliacal rising during the winter associates them strongly with the infertile season, death, and the horizon, and thus are "low," while the "northern" constellations are high in the sky during the winter and invisible at their vernal heliacal risings.

---

that the "apple doesn't fall far from the oak." Cf. Pipping's "yew cult" in the Edda-Studier and Winterbourne 76-78.

The human urge to assign directionality along Cartesian coordinates to a rotating strip of lights in the sky accords well with work in the field of "embodied logic" by George Lakoff and Mark Johnson. In their view, humans, as basically vertical beings, turn the physical experience of their own verticality into an image-schema that is subsequently mapped onto other domains: here, onto a feature of the night-time sky. This metaphorical use of verticality then brings the Milky Way into contact with other verticality metaphors, which leads to a polyvalence that includes a "vertical object as tree" metaphor. For this reason, the "Milky Way as tree" metaphor exists in numerous cultures, frequently with shamanistic overtones. Furthermore, Lakoff and Johnson's metaphor of "good is up" (23) applies to the tree schema, as the forces of evil dwell at the roots of the tree, while the less threatening bird element resides at the top.

The "Geography" of Völundarkviða

Now, finally, to the Völundarkviða. The plot can be summarized quickly: Völundr the smith and his two brothers come across three migratory swan-maidens who have settled in on a lake. The three brothers, whom the prose introduction credits with "Finnish" (i.e., Sámi) origin, marry the swan-maidens, who remain with them for eight years, but fly away in the ninth (nine units being the Eddic standard period for celestial events). The two brothers head off after the swans, one to the east, one to the south, while Völundr remains in the Wolf Dales and drowns his sorrow in obsessive smithying, with the result that he surrounds the *lind* ('linden tree') with 700 rings.[4] Here ends the swan-maiden section.

The vengeance section begins with a nocturnal visit by King Níðuðr and his men while Völundr is out hunting. They discover his 700 rings strung up at his smithy and confiscate one. Völundr returns, infers from the absence of the ring that his swan-maiden has reappeared and taken it, prepares a tasty bear steak, and falls asleep happy. He soon has a rude awakening as Níðuðr kidnaps him and carries him off to his court. At the urging of Níðuð's even more evil queen, Níðuðr hamstrings the smith, puts him out on the island Sævastaðr, and sets him to creating metallic works of art. Völundr, understandably upset at his loss of locomotion, avenges himself by killing the king's sons and turning bits of them

---

[4] Following Finnur Jónsson (*Lexicon*, "lind"), I read Vkv 5 as *lukþi hann alla − lind baugom vel* (rather than acc. pl. *bauga*), but take the *lind* to refer to the World (linden) Tree (rather than to a length of twine, as Jónsson would have it). The passage seems to imply that tying 700 rings around the old linden tree would serve as a beacon to his swan maiden and hasten her return.

into jewelry, and by seducing the princess and leaving her with child. For reasons left unexplained in this poem but elucidated in other traditions, Völundr acquires a talent for levitation and flies off after achieving his vengeance.

We begin the search for the island Sævastaðr with the provenience of the swan-maidens. The poem explicitly states – twice in the first stanza – that they arrive from the south and are southern ladies (*sunnan*; *suðroenar*). In light of the preceding discussion, their southern-ness immediately suggests two things: first, that we need to align our narrative with the celestial grid of the upright World Tree and, second, that the swan maidens, by their sudden appearance and disappearance, belong somewhere "off the map," i.e., below the horizon. This suggests the southern – or lower – part of the tree as a candidate for the general neighborhood of the action. The arrival of the three swan-maidens has a parallel in Völuspá 20's mention of the three norns who also come from under the tree (*ór þeim sæ – er und þolli stendr* 'from the lake – that stands below the fir tree'), which adds weight to the idea of locating the swan/lake complex at the visible bottom of the tree, especially when one considers the Völundarkviða's repeated phrase *ørlög drýgia* (carrying out fate), which refers rather nebulously to the actions of the three swan maidens upon arrival and departure, and which echoes the phrase *ørlög seggia* (the fate of men) connected with the three norns in the Völuspá passage. All of this makes one suspect norns in swan's clothing.

The bottom of the World Tree, i.e., the section of the Milky Way which crosses the ecliptic in the vicinity of Sagittarius and Scorpio, is marked by a significant bifurcation known as the Cygnus Rift, a name which derives from the constellation Cygnus 'the Swan,' which dominates the top, or "northern" part of the fissure. The Cygnus constellation has a north/south orientation, with the head pointing south, just as a swan maiden might appear during her return home to the netherworld. The directions in which Völund's brothers chase their respective swan wives fit the grid well; one follows his swan south, below the horizon, while the other follows the ecliptic to the east and thus also off the visible sky. As for the linked motifs of lake and maidens, if one interprets the Milky Way as a solid, the fissure itself, with its stark contrast to the surrounding, heavily populated areas, can easily appear as a lake, namely a lake under the tree.

Now that our attention has been drawn to this section of the Milky Way / World Tree, further elements fall nicely into place. Recall that Völundr remains behind after the departure of his brothers in a place called *Úlfdalir* or 'the Wolf Dales.' The explanation of this name will require a small digression.

The Germanic cultures had a well-documented interest in the winter solstice as a time of "holiness and dread" (Motz 1983, 366), an interest which later had a profound effect on Christianity and supplied us with the word "yule." In astronomical terms, the solstice season begins with the heliacal rising of the constellation which the Romans, and hence we, call Sagittarius and which lies to the southwest of the Cygnus Rift. In the Mediterranean world, the constellation traditionally represents a centaur. The existing evidence suggests that Germanic culture saw in it a different dangerous quadruped, of the canine rather than semi-equine variety. The Völuspá repeatedly evokes the image of the dog Garmr barking in front of Gnipahellir, which suggests a canine Sagittarius positioned in front of the Cygnus Rift. If Garmr and Fenrir the wolf represent two different traditions of naming the same constellation – both canines apparently rip their bonds – the cyclic nature of the Völuspá reveals itself as an astronomical cycle: the rising of Fenrir / Sagittarius heralds the beginning of the end of the old year, and the world will soon be ready for rebirth.[5] Thus, the dales at the bottom of the Cygnus Rift are indeed the Wolf Dales, dwelling place of Fenrir/Garmr, connected to the Swan constellation by the north-south axis of the Rift.[6]

Völundr has now relocated from the top of the Cygnus Rift (the swan lake) to the bottom of it (the Wolf Dales) and considerably closer to the southern district from which the swan maidens had emanated and to which they had regressed. As will have become apparent by now, the southern end of the Milky Way is decidedly the bad neighborhood, given its proximity to Hel, magical wells, and ominous norns. Immediately adjacent to the Sagittarius wolf lies the constellation Scorpio, which is threatening and reptilian in all Indo-European cosmologies (e.g., ruling the house of Death or withholding water à la Vṛtra and Indra; cf. Watkins 460-461). The Grímnismál (35), for example, describe the serpent Niðhöggr as dwelling beneath the tree and gnawing at it, thus threatening its very existence. What little iconographic evidence we have points to an identity of Niðhöggr with Scorpio. One of the Gallehus horns, for example, shows a supine serpent with open jaws attacking a tree's bifurcating roots from below, miming reasonably well the position of Scorpio below the Rift. The seventeenth-century MS AM 738 4to contains a similar image, testifying to the longevity of the tradition. To bolster the argument of the identity of Niðhöggr with Scorpio, one may point to one of the scattered *þulur*, which lists as an *Orma heiti* the word "skorpión," a rather remarkable non-native entry for the native worm or dragon, unless the author was thinking of the constellation at the time (Jónsson 1912, 675). The Icelandic designation of 'scorpion' by the word *sporðdreki* (literally a "tail-dragon") further demonstrates the Northern association of the familiar "reptile/worm" with the arthropod "scorpion" whose habitat, incidentally, stops at the northern Mediterranean.

---

[5] Puhvel calls Garmr an "allomorph" of Fenrir (220). For a related, archaeologically based interpretation, cf. Therkorn.

[6] Granted, there is in the vicinity a constellation Lupus that would make an even better candidate for Fenrir, but as it is not visible in Scandinavia and rises at the wrong time where it *is* visible, it poses an insurmountable hurdle.

16

Niðhöggr (Scorpio) and Fenrir (Sagittarius) together block the end of the Milky Way which represents the south; neither rises far above the horizon, as the ecliptic lies fairly flat at the edge of the sky when they rise in the east. They lie thus at the frontier between the positive forces of the World Tree Yggdrasill and the sub-horizontal hordes of Hel. Völuspá 39 suggests this when it connects them in the same poetic breath (in Carolyne Larrington's translation): "There Niðhögg sucks the bodies of the dead – / a wolf tears the corpses of men." Every year, the serpent and then the wolf rise, only to disappear eventually in the bright summer skies, to the relief, one imagines, of those who track such phenomena, as their sinking (cf. the final lines of Völuspá: *nú mun hon søcqvaz* 'now she will sink') signifies that the annual *ragna rök* has come full circle yet again.

Snakes and Smiths

Niðhöggr the odious serpent brings us back to Völundr and the odious king Níðuðr. Besides the striking similarity in their names, they share another association. In Völund's cameo appearance in the Anglo-Saxon *Deor* poem, one learns in line 1 that *Wēlund him be wurman – wræces cunnade* (Wēlund experienced misery with snakes). It has often been suggested that "Wurmas" was the name of evil king Nīðhād's clan. If one not unreasonably links Nīðhād, Níðuðr, and *Þiðreks saga*'s Níðungr to the Niðhögg/Scorpio complex, the reasons for his "worminess" become clear; he must, in fact, have been at some point a dragon (a "wurm"), and thus positing a clan name of "Wurmas" merely gilds the lily.

The Edda returns frequently to the relations between smiths and dragons in its treatment of the smith Reginn and the dragon Fáfnir. Fáfnir's lair is located on the Gnitaheiðr. Note the similarly between Gnipahellir in the Völuspá (as above, the cave before which Níðhögg's neighbor Garmr barks) and Gnipalundr in the first Helgi Hundingsbani poem (like Níðhöggr, also associated with snakes, albeit sea-snakes, and, like Níðhöggr, located just above the horizon). It is tempting to posit a confluence of Germanic traditions here, pointing to a heath/grove/cave with a Gnipa- or Gnita- prefix, where the Níð-prefixed dragon and the bound canine dwell at the threshold between World Tree and the destructive forces of Hel.

Reading the Völundarkviða against this astronomical background permits us to follow Völund's movements and interpret them as contrary to those of the swan maidens. Whereas the women *rise* from below the horizon and disappear again beneath it, Völundr initially *descends* into the Wolf Dales, falls into the power of the dragon king on the ecliptic, and finally rises back up north where he belongs. If any of this argument has merit, it explains why we have a conflation of swan maiden motifs with the story of Völund's vengeance. Not only do both swan maidens and smithies have the power of flight, but they occupy the same celestial lake tucked into the roots of the World Tree, just above the ecliptic. It is very tempting to see the three norns in the swan maidens and to seek their counterparts in the prominent line of three stars of Cygnus (Cygni ε, γ, and δ), but this is hardly subject to proof. In any case the bond between the two sections of the poem derives from physical proximity in the sky, not merely from the poet's "hübscher Gedanke," as Jan de Vries would have it (189). De Vries' belaboring of classical tradition – Völundr as Icarus, etc. – remains as valid as ever, but the parallels appear to derive from a common Indo-European astral tradition rather than necessarily from Germano-Roman cultural contact.

While it may seem absurd to seek to locate in the real world the scene of an Eddic tale, the contextual evidence points to complex of astral myths centered around one of the two junctures of Milky Way and ecliptic. The echoes of similar myths in other Indo-European settings suggest that the basic motifs – world tree, evil forces below the horizon, and a destructive reptile on the threshold between them – derive from a common mythological stock which assigned narrative meanings to the night-time sky, and that the island upon which Völundr suffers his captivity is located in a lake in the Cygnus Rift at the bottom of the World Tree.

Bibliography

I. Sources

Ælfric. Homilies of Ælfric: A Supplementary Collection. Volume II. Pope, John C., ed. London: Oxford University Press, 1968.

Edda: Die Lieder des Codex Regius nebst verwandten Denkmälern I: Text. Neckel, Gustav (ed.). Rev. Hans Kuhn, 5th edition. Heidelberg: Winter, 1983.

Jónsson, Finnur. Den Norsk-Islandske Skjaldedigtning. Vol I B. Copenhagen: Gyldendalske Boghandel, 1912.

The Poetic Edda. Carolyne Larrington (translator and editor). Oxford World's Classics. Oxford: Oxford University Press, 1996.

II. Secondary Literature

Aveni, Anthony F. Skywatchers. Austin, Texas: University of Texas Press, 2001.

DuBois, Thomas A.: Nordic Religions in the Viking Age. Philadelphia, University of Pennsylvania Press, 1999.

Eliade, Mircea. Shamanism: Archaic Techniques of Ecstasy. Princeton, New Jersey: Princeton University Press, 1964 [renewed 1992].

Johnson, Mark: The Body in the Mind. Chicago, University of Chicago Press, 1987.

Jónsson, Finnur: Lexicon Poeticum Antiquae Linguae Septentrionalis (Ordbog over det Norsk-Islandske Skjaldesprog). Copenhagen: Atlas, 1966.

Krupp, Edwin C.: Echoes of the Ancient Skies: The Astonomy of Lost Civilizations. Old Saybrook, Connecticut: Konecky and Konecky, 1983.

Lakoff, George and Mark Johnson: Metaphors We Live By. Chicago, University of Chicago Press, 1980.

Meller, Harald, ed.: Der geschmiedete Himmel. Stuttgart: Konrad Theiss Verlag, 2004.

Motz, Lotte: "The Cosmic Ash and other Tress of Germanic Myth." Arv 47 (1991), 127-141.

Motz, Lotte: "The Northern Heritage of Germanic Religion." The Mankind Quarterly 23 (1983), 365-382.

Pentikäinen, Juha. Saamelaiset: Pohjoisen kansan mytologia. Helsinki: Suomalaisen Kirjallisuuden Seura, 1995.

Pipping, Hugo: "Eddastudier II." Studier i nordisk filologi 17.3 (1926).

Puhvel, Jaan: Comparative Mythology. Baltimore, Maryland: The Johns Hopkins University Press, 1987.

Schele, Linda, David Freidel, and Joy Parker. Maya Cosmos: Three Thousand Years on the Shaman's Path. New York: William Morrow and Company, Inc., 1993.

Schnurbein von, Stefanie: "Shamanism in the Old Norse Tradition: A Theory between Ideological Camps." History of Religions 43 (2003) 116-138.

Therkorn, Linda L.: Landscaping the Powers of Darkness and Light: 600 BC – 350 AD Settlement Concerns of Noord-Holland, in: Wider Perspective. Dissertation, University of Amsterdam, 2004.

Vries de, Jan: "Bemerkungen zur Wielandsage", in: Schneider, Hermann, ed. Edda, Skalden, Saga: Festschrift zum 70. Geburtstag von Felix Genzmer. Heidelberg: Carl Winter Universitätsverlag, 1952.

Watkins, Calvert: How to Kill a Dragon: Aspects of Indo-European Poetics. Oxford: Oxford University Press, 1995.

Winterbourne, Anthony: When the Norns have Spoken: Time and Fate, in: Germanic Paganism. Madison, NJ: Fairleigh Dickinson University Press, 2004.

Prof. Dr. James Ogier
Department of Foreign Languages
Roanoke College
Salem, VA 24153
USA
E-Mail: ogier@roanoke.edu

Insular Myths in the *Nibelungenlied*: Was Siegfried Slain on an Island?

It seems evident that anyone who wants to give a summary of the content of the *Nibelungenlied* has to begin by explaining the meaning of the epic's title: Song of the Nibelungs. Unfortunately, the question of its meaning belongs among the greatest unsolved enigmas of the *Nibelungenlied* (J. Heinzle 2005, 29-30). One should therefore leave this topic aside and start rather from a base of undisputed historical facts about the song of the enigmatic Nibelungs[1].

Nibelungenlied is the modern title given to one of the most famous heroic epics of German literature. The three oldest manuscripts transmitting the text date from the thirteenth century.[2] They are probably copies of three different versions differing in length and certain details but not in the main structure, composed in more than 2300 stanzas totalling almost 10 000 long-lines.

The presumed original versions date from around 1200 A.D. As with the majority of German heroic epics, there is practically no evidence for the name of the author. The only thing that we know for certain is the fact that the unknown

---

[1] Jürgen Breuer provides an overview of the ancient Frankish lineage(s) of the Nibelungen, "Die historischen Nibelungen und ihre Dynastie in Geschichtsschreibung und Dichtung im Mittelalter", in: J. Breuer (ed.): Ze Lorse bi dem münster. Das Nibelungenlied (Handschrift C). Literarische Innovation und politische Zeitgeschichte. München 2006, 123-147; however, Breuer does not discuss any of the presumed mythical origins of the Nibelungen. Connections between mythical origins of the Nibelungs and heroes such as Siegfried and Brünhild clearly appear in the Scandinavian versions of the Sagas, but scholars such as Joachim Heinzle argue that these mythical elements belong to secondary features added to the Germanic legends by Nordic poets – see Joachim Heinzle: Die Nibelungen. Lied und Sage. Darmstadt 2005, 20-24 and 27. Jan-Dirk Müller, however, shows how mythical elements are first dissimulated, then reduced until, in the second part of the epic, mythical elements, especially the mythical power of the 'Nibelungs', get re-enacted to the point of causing the downfall of all heroes – J.-D. Müller: Das Nibelungenlied. Berlin 2002, 144-155. For a concise résumé of the mythical dimensions in the *Nibelungenlied* see Otfried Ehrismann: Nibelungenlied. Epoche – Werk – Wirkung. München 2002, 27-34.

[2] Since July 31st these three parchment manuscripts have been listed by UNESCO in its program "Memory of the World". The three manuscripts belong to the Bayerische Staatsbibliothek, Munich (Ms. A), the Stiftsbibliothek of St. Gallen in Switzerland (Ms. B), und die Badische Landesbibliothek in Karlsruhe (Ms. C).

22

author set down in epic form various legends with Germanic and Hunnish origins
that had been passed on orally for centuries.

I. Legends of Germanic and Hunnish Origin

Most of the action of the Middle High German Nibelungen epic centers around
the legendary downfall of the Burgundian kingdom. It is beyond doubt that this
legend can be traced back to historical events of the migration period during the
Dark Ages. In fact, the historical Burgundians belonged to the Germanic peoples
from the Baltic Sea who reached the Main and Rhine rivers in the fourth century
A.D. In the fifth century, the Burgundians expanded across the Rhine,
establishing their royal seat on the left bank in or near the city of Worms.
"However, further Burgundian expansion to the west ended in 436/ 437 with
their defeat at the hands of the West Roman general Aetius, whose Hunnish
mercenaries annihilated most of the Burgundians as well as their king, Gundahar.
The survivors were resettled in Sapaudia (Savoy) as subjects of the Roman
Empire" (Werner Wunderlich, in: The Nibelungen Tradition, p. 60).
　　Although the Hunnish leader who defeated the Burgundians was not Attila,
later historians such as Paulus Diaconus made the most powerful of all Hunnish
kings responsible for the annihilation of the Burgundians and their Rhenish
kingdom (ibid). This is roughly how and why those legendary events and their
heroes have since become part of the literary tradition of the Nibelungs.

II. Scandinavian Myths and the Nordic Nibelungen Tradition

Although specialists speak of a 'literary tradition of the Nibelungs'– as opposed
to oral traditions – they immediately admit that there are practically no traces of
relevant written legends or epics in medieval German prior to the *Nibelungenlied*. It is only through literary documents of the Scandinavian heroic tradition
that we are informed about lost German Nibelungen legends and transformations
that the legends might have undergone during the seven centuries separating the
Nibelungenlied from the historical events.
　　Oral or written memories of those Germanic and Hunnish warrior kings are
mainly preserved in such famous works of ancient Nordic literature as the Poetic
Edda, the *Volsunga saga* and the *Thidrekssaga*. Although their manuscript
tradition cannot be dated earlier than the second half of the thirteenth century,
parts of the works mentioned reflect a long and complicated merging process of
individual heroic lays or prose sagas into one story connecting the fall of the
Burgundians with the legends about Siegfried, the Dragon Slayer, and the

Valkyrie Brynhild, spellbound into sleep by her father Odin, the highest God in Nordic mythology.

At the end of this literary merging process the Nordic story reads as follows: Siegfried, called Sigurd in the Nordic tradition, is the son of the Frankish king Sigmund. Young Sigurd proves his heroic strength by killing a dragon that guards an immense treasure. After having taken possession of the treasure and some other magic objects, Sigurd awakens Brynhild from the sleep to which Odin had condemned her. Sigurd falls in love with Brynhild, but leaves her for the Burgundian court to woo the Burgundian princess, named Gudrun. In the Nordic Nibelungen tradition the Burgundians are called Niflungar, the Scandinavian version of the German name Nibelungen.

There, Gudrun's mother gives Sigurd a magic drink of forgetfulness. Sigurd forgets his betrothal to Brynhild, agrees to marry Gudrun, and even becomes the blood brother of Hogni and Gunnar, two of Gudrun's brothers. In addition, Sigurd wins Brynhild for Gunnar, who is unable to cross the wall of flame around Brynhild's hall. Sigurd disguises himself as Gunther and passes through the fire. Thus, Brynhild is successfully deceived and agrees to marry Gunnar. After the marriage of the two royal couples, Brynhild becomes unhappy because she envies her sister-in-law. When she slanderously accuses Sigurd of having broken his oaths, Gunnar and Hogni agree to murder Sigurd, a task that they ask their brother Guthorm to carry out, because he is not under oath. In most of the Nordic sagas, Guthorm does kill Sigurd, but there are different versions of the whereabouts of Sigurd's death. In some Eddic lays, Sigurd is stabbed to death, lying in bed beside Gudrun. According to other versions, Sigurd is killed outside in the open, either returning from a "Thing" meeting or in the course of a hunt. But the place is never described as an island.

After the murder, Brynhild kills herself and is burnt together with Sigurd on his funeral pyre.

Gudrun's mourning is overcome by another draught of forgetfulness. She agrees to marry the Hunnish king Attila, called Atli in the Nordic tradition. Atli invites Gunnar and Hogni to his court with the intention of procuring their treasure. Gudrun is not able to save the lives of her brothers, nor the hoard of the Nibelungs, but she finally kills Atli by burning him along with most of his men in his hall.

III. Courtly Transformations in the Middle High German Epic of the Nibelungs

In the German epic about the Nibelungs these 'old stories' are retold in a completely modernized version adapted to the courtly culture of the twelfth century as it manifested itself in courtly epics and love poetry.[3] Dominant themes such as courtly love and knighthood are reflected in various aspects of the epic: its poetic form, its narrative structure, as well as its reshaping of the main characters. The stanza form, for example, is borrowed from a German courtly poet known as the Kürenberger.[4] The narrative structure follows the model of Arthurian epics, and the ancient barbarian heroes are transformed into models of courtly kings and queens. The Burgundian court with its royal seat at Worms on the left bank of the river Rhine is renowned for its supreme honour embodied in the three kings Gunther, Gernot, and Giselher, and their sister, Princess Kriemhild. Of equally high esteem is the royal court of King Sigmund, Queen Sieglinde and crown prince Siegfried with its seat at Xanten on the Lower Rhine. Even the Hunnish ruler Etzel acts according to the new rules of courtly manners, although he still remains faithful to his heathen belief.

IV. New Roles and Realms for Siegfried and Brünhild

It therefore may not come as a surprise that the unknown poet transforms the archaic drama of the rise and fall of heroes and kingdoms into a highly dramatic love story concentrated around Kriemhild and Siegfried. In the new spirit of courtly ideals, the archaic form of a fatal love between Brünhild and Siegfried had to be replaced by a modern conception of ennobling courtly love. Consequently, the Burgundian princess is promoted to the rank of the main character because she is not only praised as the noblest princess but the most beautiful and virtuous woman in the whole world. Similarly, young Siegfried is not introduced as the dragon-slayer, but as a perfectly well-educated crown prince of the Rhenish Lowlands. The author only alludes to the heroic deeds of his youth, although in a version that differs from the Nordic Nibelungen tradition. In the Middle High German epic, Siegfried took possession of the invaluable treasure by killing the two sons of King Nibelung, thus becoming

---

[3] For a résumé of the courtly transformation of the ancient heroic elements see Ursula Schulze: Das Nibelungenlied. Stuttgart 1997, 142-176.

[4] Some *Nibelungenlied* scholars, however, defend the theory that the stanza form had been borrowed by the Kürenberger. New evidence to support this theory will be discussed by Johannes Rettelbach in a paper to be published in the conference volume "Das *Nibelungenlied* und das *Kitab Dede Korkut*". Beiträge des Ersten interkulturellen Symposiums veranstaltet von der Slawistischen Universität Baku, der Universität Mainz und der Oswald von Wolkenstein-Gesellschaft, Frankfurt/ Main, in Baku, Aserbaidschan, 30. September bis 4. Oktober 2009. Hrsg. von Hendrik Boeschoten, Sieglinde Hartmann und Uta Störmer-Caysa. Wiesbaden: Dr. L. Reichert Verlag 2010 (in preparation).

king of the Nibelungs, whose royal castle is situated on an island within a vast country far across the sea. In addition, Siegfried gained his reputation as the strongest hero of all times not simply because he had slain the dragon, but because he bathed in its blood and thus became invulnerable except for one spot between his shoulders.

In this new constellation, Brünhild's importance is reduced to a secondary role. Although the German author elevates her to the rank of a queen of Iceland, she does not act as would a courtly ruler of the civilised world. Her realm lies across the sea as far away from the courtly world as the kingdom the Nibelungs. It is called Iceland, but the poet's description evokes a country on the outskirts of the known world rather than an island in the North Sea. Consequently, Brünhild resembles more an Amazonian queen of antiquity than an ancient Germanic Valkyrie. Depicted as an invincible virgin, she would accept only the strongest suitor, one able to overcome her supernatural forces.

Several hints given by the poet suggest that Siegfried met Brünhild before getting to the Burgundian court. But the strongest hero of all times has devoted his whole heart and love service entirely to Kriemhild. He only accepts the task of overcoming Brünhild for Gunther because the Burgundian king promises him Kriemhild's hand in exchange. Although Siegfried does not deflower Brünhild, he deceives the fierce virgin queen more often and with stronger means than he does in the Nordic Nibelungen tradition. First Siegfried deceives Brünhild upon his arrival at Iceland by not wooing the maiden queen for himself but rather by pretending to serve as a liegeman of King Gunther. The two ensuing deceptions are of a different kind, because Siegfried uses his magic powers to overcome Brünhild's strength in the initial contest for the queen's hand and, later on in Worms, to submit her to Gunther in their bed chamber.

Brünhild does not seem to recognize how she has been deceived, but she shows a general suspicion. Since Gunther does not give her any further explanation of the whole matter, Brünhild's heart fills up with anger and envy. Although the poet does not depict her as motivated by unrequited love, he clearly says that the Burgundian queen has malicious intentions when she talks King Gunther into inviting Kriemhild and Siegfried to a courtly festival in Worms. Driven by her suspicion and her envy, Brünhild provokes Kriemhild, and the latter publicly accuses her of having lost her virginity to Siegfried and not to her husband. Despite an immediate, formal rehabilitation of her honour, Brünhild seeks revenge. Hagen, Gunther's vassal and a long-time opponent to Siegfried, takes on the task. He persuades Gunther, his lord, to eliminate Siegfried. Hagen deceives Kriemhild in order to find out how to wound and kill Siegfried. He arranges for a treacherous hunting party on the right bank of the river Rhine and succeeds in murdering the hero according to plan.

The second part of the epic tells of Kriemhild's bloody revenge, which causes the fall of the whole kingdom, the death of all its heroes, and the loss of the mythical Nibelungen treasure.

V. The Topographical Settings of the *Nibelungenlied*

In comparison with the Nordic Nibelungen tradition, the topographical settings of the Middle High German epic are depicted with much more attention to geography and to the details of its specific landscapes. Whereas the place names may vary in Scandinavian works or are not mentioned at all, the German poet uses names and topographical particularities in a systematic way to disclose new aspects of his main characters or even of the entire narrative.

To begin with the main centre of the action, the Burgundian court and its kingdom are named with a variation of the High German spelling that seems to allude to the French form of the name, 'Bourgogne'. Specialists suggest that we should understand this artificial form as an allusion to Beatrix of Burgundy, the second wife of Emperor Frederick I.[5] In addition, this spelling probably served as a hint to the audience, as if to say: I am telling the old stories of ancient Burgundy, but I will show you how this could happen again in our times. Consequently the whole action is transposed into a geographical setting that corresponded to the knowledge and the imagination of a twelfth century audience.

For a German audience, though, it was clear that the royal seat of the Burgundian kingdom had to remain in the city of Worms on the left bank of the Upper Rhine.

For the residence of the Hunnish court, however, the poet chose a new name, Etzelburg, which means 'Attila's castle', a name that did not correspond with any appellation transmitted in the Nordic Nibelungen tradition nor with actual Hungarian place names. The author may have chosen this neologism as a diplomatic means to avoid allusions to contemporary Hungarian cities on the River Danube and their Christian leaders. However, since the poet had the intention of characterising the Hunnish king and his kingdom as a pagan neighbour of the Christian world, he interposed Margrave Rüdiger of Pöchlarn and his border principality on the Danube between Etzel's heathen empire and the other Christian principalities on the German part of the Danube.

---

[5] Latest discussion by Heinz Thomas: "Li conte de Bourgogne – li conte de Rome. Die Staufer im *Nibelungenlied*. In: J. Breuer (ed.): Ze Lorse bi dem münster. Das Nibelungenlied (Handschrift C). Literarische Innovation und politische Zeitgeschichte. München 2006, 85-102.

VI. Brünhild's Kingdom and the Realm of the Nibelungs

Likewise, Brünhild's kingdom is transferred from the Rhineland or other parts of western Germany to a geographical outpost across the North Sea, where medieval historiography situated the Northern end of the world.[6] It is a country (not an island) called "Îslant" (= Iceland). The author may have situated Brünhild's realm as far away from the confines of the known civilised world to make it clear that Brünhild no longer represents a familiar figure of ancient legends or myths but that she embodies the mighty and fierce type of a lady as difficult to tame as the Amazonian queen of antiquity.[7] Accordingly, Brünhild's castle comprising 86 towers and three palaces with a huge hall decorated in green marble is depicted in terms of the fabulous splendours of the Oriental world, as far away from the German Rhineland as Iceland, but by no means to be confounded with the Other World of the Nibelungs such as the poet will describe it in the following chapter. In addition, Brünhild's court, the royal members of her family, her ladies-in-waiting, and her liegemen are all described as normal human beings.

Compared to Brünhild's realm, the kingdom of the Nibelungs is depicted with unmistakable features of the Other World. When Siegfried decides to fetch his vassals to reinforce their military strength at Brünhild's court, the author gives the following description of Siegfried's journey to the country of the Nibelungs:

---

[6] In medieval historiography from Germany Iceland was known and described as lying at the furthest Northern extension of the Ocean, the most influential source being written by Adam von Bremen (died ca. 1085): Magistri Adam Bremensis: Gesta Hammaburgensis ecclesiae Pontificum / Adam von Bremen: Bischofsgeschichte der Hamburger Kirche. In: Quellen des 9. und 11. Jahrhunderts zur Geschichte der Hamburgischen Kirche und des Reiches. Hrsg. von Rudolf Buchner. Neu übertragen von Werner Trillmich. Berlin 1961 (= Ausgewählte Quellen zur deutschen Geschichte des Mittelalters. Bd. XI), S. 135-503. In Adam's description Iceland is sometimes called an Island (Adam IV, 36 – identified with the ancient Thule), sometimes a country (Adam III, 72).

[7] Latest discussion by Ursula Schulze: Das Bild des Nordens bei Adam von Bremen und seine Reflexe im Nibelungenlied – Zur Dialogizität von Fremdheitsbeschreibungen. In: Jahrbuch der Oswald von Wolkenstein Gesellschaft. Sieglinde Hartmann and Ulrich Müller, editors. Vol. 16. 2006/2007, 43-57; Schulze did not take into account, that the author of the *Nibelungenlied* never speaks of an island when mentioning the name "Îslande" or Brünhild's realm, the later always being called 'country', MHG. "lant": see stanzas 339, 344, 50, 354, 360, 374, 382, 383, 384, 400, 405, 416, 419, 443, 444, 446, 475, 478, 479, 507, 512, 522, 523, 526, 527, 550 (edition by S. Grosse); apparently, the author of the *Nibelungenlied* understood the name "Îslant" in its literal sense: 'land of ice'.

"Hidden in his magic cloak, Siegfried left by the gate that opened onto the shore. There he found a bark, and, boarding it unseen, sculled it swiftly away as though the wind were blowing it. None could see the helmsman, yet, propelled by Siegfried's huge strength, the little craft made such headway that people thought a gale must be blowing it. But no, it was being sculled by Siegfried, fair Sieglind's son.

After voyaging for the rest of the day and on into the night for upwards of a hundred miles, Siegfried thanks to his vast exertions reached a land that was named after the Nibelungs, of whose great treasure he was lord. Alone though he was, the dauntless warrior sailed to a large island in the river (MHG "Der helt fuor aleine ûf einen wert vil breit", stanza 485, 1), quickly moored the bark, and made for a castle on the hill in quest of shelter (…)." (The Nibelungenlied. Translated by Arthur Thomas Hatto. London: Penguin Books 1969, p. 70)

This passage reads as follows in the three oldest manuscripts of the *Song of the Nibelungs*:

Manuscript A (Munich, Bayerische Staatsbibliothek, Cgm. 34, stanza 453-454)

"Bi des tages cite	vnd bi der einen naht
Kom er zeime lande	mit michelr kraft,
hvndert langer raste	vnd dannoch lihte baz;
daz hiez Niblvnge,	*da er den grozen hort besaz.*

Der helt fûr alleine	*vf einen wert bereit.*
Das schif geband vil balde	der riter vil gemeit.
Er gie zů eime berge,	darufe ein burch stůnt,
vnd suehte herberge,	so die wegemude tůont. "[8]

Manuscript B (St. Gallen, Stiftsbibliothek, Codex 857, stanza 482-483)

"Bî des tages zîten	und in der einen naht
kom *er zeinem lande*	mit grœzlicher maht,
wol hundert langer raste	und dannoch baz.
daz hiezen Nibelunge,	*dâ er den grôzen schatz besaz.*

Der helt, der fuor aleine	*ûf einen wert vil breit.*
daz schiff gebant vil balde	der ritter vil gemeit.
Do gie er zuo einem berge,	dar ûf ein burc stuont, (…)."[9]

---

[8] All quotations from manuscript A are taken from the edition by Michael S. Batts: *Das Nibelungenlied*, critical edition, Tübingen: M. Niemeyer 1971.

[9] All quotations from manuscript B are taken from the edition by Hermann Reichert (ed.):

Manuscript C (Karlsruhe, Badische Landesbibliothek, Codex Donaueschingen 63, stanzas 495-496)

"Bi des tages cite	und in der einen naht
chom *er zeinem lande*	mit grœzlicher maht,
daz hiez zen Nibelungen	und waren sine man.
Lant unde burge,	daz was im allez undertan.

Der herre fuor aleine	*uf einen wert vil breit.*
Daz schiff gebant vil balde	der ritter vil gemeit.
Do gie er zeinem berge,	da ein burch stuont. (...)".[10]

It is important to note that the three oldest manuscripts do not differ in highlighting the land of the Nibelungs as a realm of the Other World where the castle and the treasure of the Nibelungs are guarded by monstrous creatures of the Other World.

Consequently Siegfried finds the gate of the castle guarded by a giant. But the ferocious gatekeeper does not recognize his liege lord. Therefore, Siegfried has to overcome him in a fierce fight. The din of the savage fighting is not only heard inside the whole castle but also far away through a mountain in which the 'fearless' dwarf Alberich guards the Nibelungen hoard. Again, Siegfried has to defeat his own Lord Treasurer to be duly recognized as the sovereign lord of all Nibelungs. The dwarf then assembles 3000 of their warriors in the hall of the Nibelungs, of whom a thousand of the best are chosen to sail back with Siegfried to Iceland.

Later on, after the marriage of Siegfried and Kriemhild, the poet tells his audience that the royal couple was residing in a northern country, called Norway in manuscript A and B. That is the place to which Gunther sends his messengers with the fateful invitation to the festival at the Burgundian court. The Burgundian messengers need three weeks (12 days in MS C) to reach this castle of the Nibelungs (MHG. "ze Nibelungens burge"); "and there at Nibelung's stronghold" (Hatto, ibid., p. 101) they find the addressee of the invitation.

Nibelung specialists do not have any convincing explanation of why the royal couple moved from the Lower Rhine to that Scandinavian country, which, on top of it all, seems to be reached by land and not by sea. The great Hatto, whose beautiful English translation I am citing from, suggests that the "poet has merged

---

Das Nibelungenlied. Nach der St Galler Handschrift. Berlin / New York 2005.

[10] All quotations from manuscript C are taken from the edition by Ursula Schulze (ed.): Das Nibelungenlied. Nach der Handschrift C der Badischen Landesbibliothek Karlsruhe. Düsseldorf / Zürich: Artemis & Winkler 2005.

the Netherlands, Norway, and the Nibelungenland into one indeterminate northern realm ruled by Siegfried" (ibid. p. 101).

Whatever the reasons for this geographical shift, it remains to be noted that the topography of the royal seat in Norway is not described. The author does not give any hint about the Nibelungen hoard nor does he make any allusion to an island.

VII. The Place of Siegfried's Death: A 'Spacious Isle in the River' Rhine

The only other island to be chosen as scenery by the poet of the *Nibelungenlied* is the isle in the river Rhine where Siegfried is slain. Strangely enough, this striking parallel has not yet found the attention it deserves, especially since the poet uses exactly the same words to allude to it: "ûf einen wert vil breit", which means: on a large island.

Therefore, one must ask now: why was Siegfried slain on an island?

Indeed, the poet does not leave any doubt about his scenery, for he mentions the island twice in that part of the epic. First he tells us that, after having left the royal seat of Worms on the left bank of the Rhine and after having crossed the river, that the

> "proud and intrepid hunters were told to set up their lodges on a spacious isle in the river on which they were to hunt, at the skirt of the greenwood over towards the spot where the game would have to break cover. Siegfried, too, had arrived there, and this was reported to the King. Thereupon the sportsmen everywhere manned their relays." (ibid. chapter 16, p.125);

MS A stanza 871:
> "Si hiezen herbergen   fur den grûnen walt
> Gen des wildes abelaufe,   die stolzen iægere balt,
> da si da iagen solden,   *vf einen wert vil breit.*
> Do was ŏch komen Sifrit;   daz wart dem künige geseit."

MS B stanza 925:
> "Si hiezen herbergen   für den grüenen walt
> gegen des wildes abloufe,   die stolzen jeger balt,
> dâ si dâ jagen solden,   *ûf einen wert vil breit.*
> dô was ouch komen Sîvrit.   daz wart dem künege geseit. "

MS C stanza 936:

> "Si hiezen herbergen fur den grüenen walt
> gen des wildes abeloufe, diem stolzen jägere balt,
> da si da jagn solden, *uf einen wert vil breit.*
> do chom der herre Sivrit. daz wart dem kunige geseit"

When the hunt is over and it comes to the final act of treason, the poet reminds his public that the scene is still set on the same island: Though mortally wounded, Siegfried strikes his murderer Hagen so powerfully that the traitor falls "reeling under the weight of the blow and the isle echoed loudly". (Hatto translates the MHG "wert" with 'riverside', ibid. p. 131; MHG original: MSS A, B and C almost identically read "von des slages krefte der wert vil lût' erhal", stanza 986, 2, edition by S. Grosse).

Modern interpreters of the Nibelungen epic easily overlook this topographical setting.[11] One of the reasons could be that the actual valley of the Upper Rhine no longer resembles the landscape as it appeared from medieval times on up to the nineteenth century. Since 1817, the year the engineer Johann Gottfried Tulla (1770-1828) started the gigantic task of cutting a straight waterbed through the numerous meanders of the Upper Rhine, the characteristic features of its landscape, including hundreds of bigger and smaller islands, began to disappear.

The poet's scenery therefore refers to topographical elements that actually did correspond with the characteristics of the right bank of the medieval river Rhine as it can easily be reconstructed for the region across from the city of Worms.

But there is still another reason why modern interpreters overlook this realistic landscape feature in the description of Siegfried's murder. The rest of the scenery, the fauna with its wild beasts such as lions and bisons, uncommon for the Rhineland, as well as the flora of a highly stylized *locus amoenus* rather

---

[11] The *opinio communis* pretends that Siegfried was slain in the Odenwald, a middle mountain situated on the right bank of the Rhine across from Worms. In fact, manuscript C mentions that Siegfried proposed to ride across the Rhine to go hunting in the "Otenwalde" (stanza 919,3 – ms. A, stanza 854,3, and B, stanza 908,3, mention the Alsatian mountains of the Vosges). However, these placenames are not taken up in the following chapter with the description of Siegfried's murder. All three manuscripts repeat that the hunting forest (without name) lies on a 'large island' within the river Rhine. In addition, in ms. C, there is a commentary added where it is explained that Siegfried had been slain "vor dem Otenwalde" in a village called "Otenhaim" (stanza 1013). Indeed old pictures of the plain between the Odenwald and Worms show that this village was situated on an island within the river Rhine.

give the impression of a place beyond the real world where a purely symbolic meaning seems to prevail.

This symbolism has long been recognized, for the symbolic meaning of the hunting scene also becomes clear through the narrative structure of this chapter.

The poet describes the murder in three scenes, following a poetic strategy that shows Siegfried first in the role of a hunter with unsurpassable fortitude and then in the role of the guileless and therefore defenceless prey of his two murderous hunters, King Gunther and his vassal Hagen. Between the sequences of these contrasting scenes a comic play with a wild bear is inserted to relax the tension before the final tragedy. Thus the episode shows us the two sides of Siegfried's heroic character as if in two large close-ups: the victorious hunter embodying the archaic type of a hero characterised by his extraordinary physical strength, and the guileless victim of envy and treason characterised as a new saint-like type of a hero whose unsurpassable interior fortitude enables him to suffer the bloodiest martyrdom for his outstanding virtues.

This new heroic Siegfried no longer needs the help of magic powers, dwarves, or giants. The place of his apotheosis does not lie on an island far beyond his Rhenish realm. The isle of the river Rhine, where the poet of the *Nibelungenlied* places Siegfried's martyr-like death, belongs to the real part of his world.

If there is a symbolic meaning to the island within the Rhine, one could come to the following conclusion: the isle where Siegfried was slain should not be confounded with an island of the Other World, such as Avalon for example, where another immortal hero of the European Middle Ages found his apotheosis (M. Oergel, 1998). But the island of Siegfried's death surely marks a decisive element of his new mythical dimension – a genuinely medieval dimension.

Bibliography

I. Sources and English Translations

Edition quoted: *Das Nibelungenlied*. Mittelhochdeutsch / Neuhochdeutsch. Nach dem Text von Karl Bartsch und Helmut de Boor. Ins Neuhochdeutsche übersetzt und kommentiert von Siegfried Grosse. Stuttgart: Reclam 2002.

Michael S. Batts: *Das Nibelungenlied*, critical edition, Tübingen: M. Niemeyer 1971 (contains the versions of the 3 oldest manuscripts A, B, and C).

The Nibelungenlied, The Older Lay of Hildebrand, and other works. Ed. by Francis G. Gentry et al. New York: Continuum 1995.

Schulze, Ursula: *Das Nibelungenlied* (based on manuscript C). Düsseldorf / Zürich: Artemis & Winkler 2005.

Hermann Reichert (ed.): *Das Nibelungenlied*. Nach der St Galler Handschrift. Berlin / New York 2005 (edition of mansucript B).

English translation: *The Nibelungenlied*. Translated by Arthur Thomas Hatto. London: Penguin Books 1969.

Burton Raffel, *Das Nibelungenlied*, new translation. Foreword by Michael Dirda. Introduction by Edward R. Haymes. Yale University Press 2006.

The Nibelungenlied. The Lay of the Nibelungs. Translated with an Introduction and Notes by Cyril Edwards. Oxford 2010.

The Nordic Nibelungen Tradition and its Main Sources

1) Poetic Edda, Iceland, ca. 1250; Ursula Dronke (ed. and trans.) *The Poetic Edda*. Vol. 1 of Heroic Poems. Oxford 1969; German translation: *Die Götter- und Heldenlieder der Älteren Edda*. Übersetzt, kommentiert und hrsg. von Arnulf Krause. Stuttgart: Reclam 2004.

2) The Prose Edda of Snorri Sturluson (1179-1241); Anthony Faulkes (Ed.): *Edda by Snorri Sturluson: Prologue and Gylfaginning* (Oxford 1982) and *Edda by Snorri Sturluson: Skáldskaparmál*, 2 vols. (London 1998); *Snorri Sturluso. The Prose Edda*. Translated with an Introduction and Notes by Jesse L. Byock. London/New York 2005.

3) The Völsungasaga, Norway ca. 1250; R. G. Finch (ed., transl.) *The Saga of the Volsungs*. London 1965; German translation: *Nordische Nibelungen*. Die

Sagas von den Völsungen, von Ragnar Lodbrok und Hrolf Kraki. Aus dem Altnordischen übertragen von Paul Hermann. Hrsg. von Ulf Diederichs. Köln 1985.

4) Thidrekssaga, Norway, ca. 1250; Henrik Bertelsen (ed.): *Thidreks Saga af Bern*. 2 vols. Copenhagen 1905-1911; German translation: *Die Geschichte Thidreks von Bern*. Übertragen von Fine Erichsen. Jena 1924. Nachdruck München 1996.

II. Secondary Literature

Except for some recent articles, we refer to the latest monographs, and conference or exposition volumes for further bibliographical references.

Boeschoten, Hendrik, Hartmann, Sieglinde and Störmer-Caysa, Uta (editors): Das *Nibelungenlied* und das *Kitab Dede Korkut*. Beiträge des Ersten interkulturellen Symposiums veranstaltet von der Slawistischen Universität Baku, der Universität Mainz und der Oswald von Wolkenstein-Gesellschaft, Frankfurt/Main, in Baku, Aserbaidschan, 30. September bis 4. Oktober 2009. Wiesbaden: Dr. L. Reichert Verlag 2010 (in preparation).

Breuer, Jürgen (editor): Ze Lorse bi dem münster. Das Nibelungenlied (Handschrift C). Literarische Innovation und politische Zeitgeschichte. München 2006,

Ehrismann, Otfrid: Nibelungenlied. Epoche - Werk - Wirkung. 2. Aufl. München 2002.

Fasbender, Christoph (editor): Nibelungenlied und Nibelungenklage. Neue Wege der Forschung. Darmstadt 2005.

Gentry, Francis G., McConnell, Winder, Müller, Ulrich and Wunderlich, Werner (editors): The Nibelungen Tradition. An Encyclopedia. New York and London 2002.

Haymes, Edward: Das Nibelungenlied. Geschichte und Interpretation. München 1999 [= UTB 2070].

Heinzle, Joachim, Klein, Klaus and Obhof, Ulrike (editors): Die Nibelungen. Sage – Epos – Mythos. Wiesbaden 2003.

Heinzle, Joachim: Die Nibelungen. Lied und Sage. Darmstadt 2005.

Herman, Nadine I.: The development of the Nibelungen-legend in various periods of German literature. Diss. Ottawa 2001.

Jefferis, Sibylle (editor): The Nibelungenlied: Genesis, Interpretation, Reception (Kalamazoo Papers 1997-2005). Göppingen: Kümmerle 2006.

McConnell, Winder (editor): A Companion to the Nibelungenlied. Columbia 1998.

Müller, Jan-Dirk: Das Nibelungenlied. Berlin 2002 [= Klassiker-Lektüren 2].

Oergel, Maike: The return of King Arthur and the Nibelungen: national Myth in nineteenth-century English and German literature. Berlin u.a.: De Gruyter 1998.

Schaefer, Ursula and Spielmann, Edda (editors): Varieties and Consequences of Literacy and Oralty / Formen und Folgen von Schriftlichkeit und Mündlichkeit. F. H. Bäuml zum 75. Geburtstag. Tübingen 2001.

Schulze, Ursula: Das Nibelungenlied. Stuttgart 1997 (= Reclam Literaturstudium 17604).

Prof. Dr. Sieglinde Hartmann
Universität Würzburg
Institut für deutsche Philologie
Am Hubland
D - 97074 Würzburg
sieglinde.hartmann@germanistik.uni-wuerzburg.de

Holy Islands and Their Christianization in Medieval Prussia

The *Chronica* of Peter of Dusburg on Prussians and their Faith

Prutheni notitiam dei non habuerunt. Et quia simplices fuerunt, ratione comprehendere eum non potuerunt,[et quia literas non habuerunt], immo nec in scripturis ipsum speculari non poterant – wrote Peter of Dusburg, a Teutonic monk, in his *Chronica Terrae Prussiae* of 1326.

He continued:

Mirabantur ultra modum in primitivo, quod quis absenti intentionem suam potuit per literas explicare. Et quod sic Deum non cognaverunt, ideo contigit, quod errando omnem creaturam pro deo coluerunt, scilicet solem, lunam et stellas, tonitrua, volatilia, quadrupedia etiam, usque ad bufonem. Habuerunt eciam lucos, campos et aquas sacras, sic quod secare aut agros colere vel piscari ausi non fuerant in eisdem . Fuit autem in medio nationis huius perversae, scilicet in Nadrouia, locus quidam dictus Romow, trahens nomen suum a Roma, in quo habitabat quidam dictus Criwe, quem colebant pro papa, quia, sicut dominus papa regit universalem ecclesiam fidelium, ita ad istius nutum seu mandatum non solum gentes predicte, sed et Lethowini et aliae nationes Liuoniae terare regebantur. Tantae enim fuit autoritatis, quod non solum ipse vel aliquis de sanguine suo, verum et nuncius cum baculo suo vel alio signo certo transiens terminos infidelium predictorum a regibus et nobilibus et communi populo in magna reverentia haberetur. Fovebat etiam prout in lege veteri iugem ignem. Prutheni resurrectionem carnis credebant, non tamen, ut debebant. Credebant enim, si nobilis vel ignobilis, dives vel pauper, potens vel impotens esset in hac vita, ita post resurrectionem in vita futura. Unde contingebat, quod cum nobilibus mortuis arma, equi, servi et ancillae, vestes, canes venatici, aves rapaces, et alia, quae spectant ad militiam, urerentur. Cum ignobilibus comburebatur id, quod ad officium suum spectabat. Credebant, quod res exustae cum eis resurgerent, et servirent sicut prius. Circa istos mortuos talis fuit illusio diaboli, quod cum parentes defuncti ad dictam Criwe venirent querentes, utrum tali die vel nocte vidisset aliquem domum suam transire, ille Criwe et dispositionem mortui in vestibus, armis, equis et familia sine haesitatione aliqua ostendebat, et ad maiorem certitudinem ait, quod in superliminari domus suae talem fixuram cum lancea vel instrumento alio dereliquit. Post victoriam diis suis victimam obtulerunt, et omnium eorum, que ratione victorie consecuti erant, tertiam partem dicto Criwe praesentaverunt, qui combussit talia. Nunc autem Lethowini et alii illarum partium infideles dictam victimam in aliquo loco sacro secundum eorum ritum comburrunt, sed antequam equi comburrerunt, cursu fatigantur in tantum, quod vix possunt stare supra pedes suos. Prutheni raro aliquod factum notabile inchoabant, nisi prius missa sorte secundum ritum ipsorum a diis suis, utrum bene vel male debeat eis succedere, sciscitarentur. Vestes superfluas aut pretiosas non curabant, nec adhuc curant; sicut quas hodie ipsas exuit, ita cras iduit, non attendens, si sint transversae. Molli stratu et cibo delicato non utuntur. Pro potu habent simplicem aquam et mellicratum seu medonem, et lac equarum, quod lac quondam non biberunt, nisi prius sanctificarentur. Alium potum antiquis temporibus non noverunt. Hospitibus suis omnem humanitatem, quam possunt, ostendunt, nec sunt in domo sua esculenta vel poculenta, quae non communicent eis illa vice. Non videtur ipsis, quod hospites bene procuraverint, si non usque ad ebrietatem sumpserint potum suum. Habent in consuetudine, quod in potationibus suis ad eaquales et immoderatos haustos se obligant, unde contingit, quod singuli domestici hospiti suo certam mensuram potus offerunt sub hiis pactis, quod, postquam ipsi ebiberint, et ipse hospes tantundem evacuet ebibendo, et talis oblatio potus totiens reiteratur, quousque hospes cum domesticis, uxor cum marito, filius cum filia omnes inebriantur. Secundum antiquam consuetudinem hoc habent Prutheni adhuc in usu, quod uxores suas emunt pro certa pecunie summa. Unde vir servat uxorem sicut ancillam, nec cum eo comedit in mensa et singulis diebus domesticorum et hospitum lavat pedes. Nullus inter eos permittitur mendicare, libere vagatur egenus inter eos de domo ad domum, et sine verecundia comedit, quando placet. Si homicidium committitur inter eos, nulla potest composicio intervenire, nisi prius ille homicida vel propinquus eius ab occisi parentibus occidatur. Quando ex inopinato rerum eventu aliquam immoderatam incurrerunt turbationem, se ipsos occidere consueverunt. Distinctionem dierum non habuerunt aut discretionem. Unde contingit, quando inter se vel ipsi cum alienis aliquod placitum vel parlamentum volunt servare, datur certus numerus dierum, quo facto quilibet eorum prima die facit unum signum in aliquo ligno vel nodum in corrigia aut zona. Secunda die addit iterum secundum signum, et sic de singulis, quousque perveniat ad illum diem, quo tractatus huiusmodi est habendus. Aliqui omni die balneis utebantur ob reverentiam deorum suorum, aliqui balnea penitus detestabantur. Mulieres et viri solebant nere, alqui linea, alii lanea, prout credebant diis suis complacere. Aliqui equos nigros, quidam albos vel alterius coloris propter deos suos non audebant aliqualiter equitare.[1]

---

[1] Jarosław Wenta, Sławomir Wyszomirski (ed.), Piotr z Dusburga, Kronika ziemi pruskiej, Monumenta Poloniae Historica, n. s., t. XIII, Kraków 2007, 52-54 (III,5).

Translation:

"The Prussians had no knowledge of God. Because they were simple people, they could not comprehend Him with their minds and, because they were illiterate, they could not perceive Him in the Scriptures. They were amazed beyond all measure that one could convey one's message to an absent person by means of letters. So because they did not know God, they therefore worshipped, erroneously, every creature as a god, that is, the sun, the moon and stars, thunder, birds, even quadrupeds, right down to the toad. They also had woods, fields, and holy waters that were so sacred to them that they dared not chop firewood nor cultivate the fields nor fish in them. Furthermore, there was in the middle of these infidels' nation, i.e., in Nadrowia, a place called Romow – deriving its name from Rome – in which lived a man called the Criwe whom they worshipped as a pope, for just as the Pope reigns over the entire Church of believers, this one ruled, by gesture or command, not only over the aforementioned Prussians, but also over the Lithuanians and the other nations of Livonia. Such was his authority that not only he himself, but any other of his clan_–_truly, even a messenger with his staff or some other sign – travelling through the lands of the aforementioned infidels would be greatly revered by kings and nobles and the common people. He also tended – as in the Old Testament – the eternal flame. The Prussians believed in the resurrection of the body, but not in way they should have; for they believed that if one is noble or peasant, rich or poor, powerful or powerless in this life, one would be so in one's further life after resurrection. For this reason they burned, along with their dead nobles, their weapons, horses, serving men and women, clothing, hunting dogs and birds, and other things appertaining to the warrior's status. With the peasants they burned whatever belonged to their service. They believed that the burned items would be resurrected along with them and would serve them as they had before. Such was the state of their diabolical delusion with regard to these dead ones that when the relatives of the deceased came to Pope Criwe and asked whether he had seen anyone pass by his house, this Criwe would describe without any hesitation the appearance of the deceased as to clothing, arms, horses, and servants, and showed as further proof the mark of a lance or other instrument in his doorjamb that the deceased had left there. After a victory they offered a sacrifice to their gods and they presented a third of everything they had gained through their victory to the Criwe, who then burned it. Now, however, the Lithuanians and the other infidels of those areas immolate the victim in some sacred placea according to their rites, but before the horses are burned, they run them until the horses are so fatigued that they can hardly stand on their legs. The Prussians rarely began anything important without first casting lots according to their rites to find out from their gods whether the thing would turn out well or badly. They did not care about having many or expensive clothes, nor do they today; the way they take them off today is the way they put them on tomorrow, not worrying if the clothes are on backwards. A soft resting-place and fine foods are unknown to them. To drink they have simple water and a honey drink or mead, and mare's milk, which back then they would not drink until it had been blessed. They did not know any other drink in olden times. They demonstrate all the hospitality they can to their guests; there is nothing edible or potable in their house that they do not share with them. They do not think that their guests have been well seen to if the guests do not drink until they are inebriated. They have the custom of obliging each other at drinking parties to consume an equal and immoderate amount, which results in members of the household offering their guest a certain measure of drink under the condition that, after they have finished their drink, the guest must also empty his glass with the same amount; and this offer of a drink is repeated until guest and hosts, wife and husband, son and daughter, are all inebriated. Following their ancient custom, the Prussians still have the habit of buying their wives for a certain sum of money. Then he keeps her like a servant; she does not eat with him at table and on certain days she washes the feet of the household members and of the guests. None of them is allowed to beg; the poor go freely among them from house to house and eat without shame when it pleases them. If a murder is committed among them, no settlement can be made until the murderer or one of his family members is killed by the dead man's relatives. When an unexpected event causes them extreme distress, they tend to kill themselves. The do not distinguish between the days of the week. This means that, when they want to have a meeting or a conference among themselves or with strangers, a certain number of days is agreed on; having done that, on the first day one of them makes a mark in some piece of wood or a knot in a strap or belt. The next day he adds a second sign, and so on each day until he arrives at the day on which the scheduled conference is to take place. Some bathed every day out of reverence for their gods; others completely detested bathing. Men and women usually spun thread, some linen and others wool, whichever they thought would please their gods. For fear of their gods, some dared not ride black horses, others white or differently colored ones." [Translation J. Ogier].

The *Chronica* of Peter of Dusburg usually constitutes a starting point for research on the religion and social order of Prussians before and during the conquest. Even though the religious system of pagan Prussia has been extensively, albeit sometimes ineffectively described in a whole range of dissertations, we are still far from formulating an accurate description of it.[2]

---

[2] Benno Winckler, Romowe in Warmien, Zeitschrift für die Geschichte und Altertumskunde Ermlands, 3 (1866), 521-526; Eduard Gigas, Der sogenante Potrimpos

41

To recapitulate Peter's text, the Holy Grove, the place where the high priest (*Kriwe*) dwelt and where offerings were made, was called the Ramowe. The Ramowe could be any place with an oak with three large and intricately twisted branches. The intricate shape of the tree guaranteed the presence of a deity, and it was at this tree that the offering altar was erected. The Kriwe was usually surrounded by a group of priests and priestesses (*rykajoth*). Despite women's low social position, the priestesses enjoyed a surprisingly high status in particular circles (*lauks*).

The Kriwe's death had a supernatural dimension. The dying priest mounted the offering pyre, called everyone to prayer and confession of iniquities, and died in the flames. Every settlement had its Ramowe. Priests and priestesses did not enter into marriage because of the knowledge that the gods could always call them to sacrifice their own lives to save the community. Their duties also included making less significant offerings; teaching religion; instructing the community on how their gods wanted them to live; blessing people; performing augury; announcing the times of holiday, harvest, or driving cattle to and from pasture; telling the dying about the delights of the afterlife; and treating the sick and the wounded. Depending on the function they performed, priests and priestesses were divided into various groups: *tuliuzs* and *liguzs* were present at a sick person's deathbed and conducted funeral rites; *burts* performed augury with wax; *seitons* selected amulets; *swalgons* arranged marriages.[3] People gathered in holy forests after achieving a victory or taking spoils, for personal reasons

zu Christburg, der sogenante Bartel und die Gustabalde zu Bartenstein, Zeitschrift des historischen Vereins für den Regierungsbezirk Marienwerder, 2 (1877), 43-54; Aleksander Brückner, Die Galindensage, Archiv für slawische Philologie, 21 (1899), 22-27; Wilhelm Friederici, Über die Lage Romow's oder Romowe's des Oberpriestersitz im heidnischen Preussen, Altpreussische Monatsschrift, 13 (1876), 227-253; Adolf Rogge, Der preussische Landberg, das älteste Romowe, Altpreussische Monatsschrift, 14 (1877), 585-592; Joseph Bender, Zur altpreussischen Mythologie und Sittengeschichte, Alpreussische Monatsschrift, 2 (1865), 577-603, 694-717, 4 (1867), 1-27, 97-135; Hans Crome, Die Religion der alten Preußen, Alt Preußen, 4 (1939), 50-54; Aleksander Brückner, Starożytna Litwa, Ludy i bogi: szkice historyczne, Olsztyn 1979; Franciszek Bujak, Dwa bóstwa prusko-litewskie "Kurche" i "Okkopirnus", Lwów 1924; Haralds Biezais, Die Religionsquellen der baltischen Völker und die Ergebnisse der bisherigen Forschungen, Uppsala 1954, s.65-128; Jan Powierski, Bogini Kurko i niektóre aspekty społeczno-gospodarcze wierzeń pruskich, Prace Komisji Historycznej Bydgoskiego Towarzystwa Naukowego, 11 (1975), 3-23; Jerzy Marek Łapo, Czcij praojca swego: o kulcie przodków w kręgu zachodniobałtyjskim na początku I tys. n.e., Popiół i kość: Funeralia lednickie, 4 (2002), 321-337. Jerzy Suchocki, Mtologia bałtyjska, Warszawa 1991. Grzegorz Białuński, Las w wierzeniach Prusów i Jadźwingów, Feste Boyen, 1 (2002), 31-42.

[3] Łucja Okulicz-Kozaryn, Dzieje Prusów, Wrocław 1997, pp. 217-331.

(misfortunes, illnesses, funerals), and during annual holidays (the beginning of spring, harvest, all souls' days [*sermen*]). The feasts involved animal sacrifice. On all souls' days the feasts took place in forests and were devoted to *Veles*, which most probably meant the deceased souls.

All attempts at a more profound interpretation of the Prussian religious system are hindered by the late origin of the sources – such as the chronicle of Simon Grunau, a Dominican from Elblag, from the beginning of the sixteenth century, the time when the last evidence of the pagan cult in Prussia began to disappear – and by the necessity of relying on references to the situation in Lithuania and Livonia.

By imposing on the Ramowe a cosmic dimension, a model of the world derived from the works of specialists in religious studies, the following schema prevents us from exceeding the bounds of current academic methodology.

It is assumed that the holy oak signified three spheres:
- the underworld, whose symbol consisted of three skulls: one each of a human, a horse, and a cow. The skulls were subject to the power of the god of death, Patollus;
- the earthly world of Patrimpus, depicted by a snake in a pot of milk covered with ears of grain;
- the sky of Perkunis, represented by the eternal flame in the Ramowe.

Female cults were connected with the lime tree, and we know that the goddess Curche may have had a female nature. The Lithuanian goddess of earth and harvest, Zeminele, was also female. The male Pushayta protected groves and forests, and lived under the lilac. Another male god, Medinis, also lived in the forest and was responsible for the swoosh of the wind in the trees, the diversity of animals, and the abundance of forest fruit. In winter, goats were a common sacrifice for Medinis, and the god appeared as a wolf to consume the offering. Sacrificial animals had to be black, white, or chestnut in colour. Deities were also offered eggs, fats, drinks, and, under the Teutonic rule, money. Some of the animals offered were not burnt but rather killed at the foot of the holy tree so that the blood would fertilize the soil. Prussians also made sacrifices of humans, especially children (little girls), who were offered to Potrimpus as an act of thanksgiving or as a prayer for grace, protection, happiness, prosperity or fertility. Since Lithuanians had their own notions of gods, it is possible that the Prussians did, too[4].

---

[4] Białuński 2002, 31-42.

Christianization in Prussia under Teutonic Rule

However, under no circumstances may the above sketch be treated as a picture of the Prussian religion before the conquest. It should rather constitute a reference point for a description of the religious situation and theological notions in Prussia under Teutonic rule. It should be noted that the conquest was not tantamount to Christianization; consequently, Christianization itself did not mean renouncing the old beliefs or accepting the new system of values.

For example, the old system was reflected in the names of areas that could be associated with cultic places and pagan holy sites. Sometimes these names were made to sound German or translated into German. Thus, *Alk, Alkeyne, Alkehnen,* and *Alkinchen* signified a holy grove, which formerly was referred to as *alka, medis* or *ramavan,* whereas *Kawken* meant the place of the presence of a devil, a gnome, or a dwarf. Medenau should be associated with a holy forest, Kurche-lauk comes from the goddess Curche, Laune-grabis designates a fairy godmother or a witch, while Perkune and Potollen are the names of concrete deities.[5]

Interestingly enough, the holy places in lime groves that belonged to the Prussian goddess Cruche became the places of the cult of Virgin Mary.[6] Because of her miraculous interventions, the Virgin Mary is regularly mentioned in the *Chronica* of Peter of Dusburg.[7]

The sacralization of names constituted a means of creating a new dominant religious realm in Prussia. This is how Peter of Dusburg described the construction of the future capital of the Teutonic Order:

Anno MCCLXXX castrum Santirii mutato nomine et loco translatum fuit ad eum locum, ubi nunc situm est, et vocatum nomen eius Mergenburgk i. e. castrum sanctae Marie, ad cuius laudem et gloriam haec translatio facta fuit.[8]

---

[5] Georg Gerulis, Die altpreußischen Ortsnamen, Berlin und Leipzig 1922, p. 234; Rozalia Przybytek, Ortsnamen baltischer Herkunft im südlichen Teil Ostpreußen, Stuttgart 1993; Grasilda Blaziene, Die baltischen Ortsnamen in Samland, Stuttgart 2000.

[6] Białuński 2002, 40.

[7] Marian Dygo, O kulcie maryjnym w Prusach krzyżackich w XIV-XV w., Zapiski historyczne, 52 (87), 237-268.

[8] "In the year 1280 the fortress of Santiri was renamed and moved to the place where it now stands and was called Mergenburg, i.e., the fortress of Holy Mary, to whose praise and glory the move was made." [Translation J. Ogier]. Wenta 2007, 181 (III, 208). Bernhard Schmid, Die Gründung der Marienburg, Altpreußische Forschungen, 6 (1929), 191-200; Jan Powierski, Chronologia początków Malborka, Zapiski Historyczne, 44 (1979), 181-207.

Holy Islands in Prussia: The Example of Marienwerder

It is time we discussed the Holy Islands. According to Peter of Dusburg:

Postquam hec castra per Dei gratiam essent eadificata, et vetus fermentum malitiae et nequitiae infidelium esset a terrae Colemensis finibus expurgatum, ut prosperum iter faceret nobis Deus salutarium nostrorum ad terras gentium vicinarum, et aliqui aestimant pro certo, magister et fratres, praeparatis iis, que ad edificationem castrorum necessaria sunt, secrete venerunt navigio ad insulam de Quidino quasi ex opposito nunc Insule Sanctae Mariae, et ibi anno Domini 1233 erexerunt in quodam tumulo castrum vocantes illud Insulam Sanctae Mariae. Sed dum vir nobilis et miles strenuus in armis de Saxonia burgrabius de Megdeburgk, dictus cum parva manu, multa stipatus militia et armigeris veniret ad castrum Colmen, intra annum, quo ibidem mansit, ivit cum magistro et fratribus, et castrum Insule Sanctae Mariae predictum transtulit de insula Quidini ad locum, ubi nunc est situm, in territorio Pomesaniae dicto Risen mutantes locum et non nomen.[9]

"After the said fortresses had been built by the grace of God and the old brew of pagan evil and iniquity had been expunged from the bounds of the land of Culm, so that the God of our salvation might make us an easy path to the lands of the neighboring heathens, as many firmly believe, with the master and the brothers having prepared everything necessary for building fortresses, they secretly shipped over to the island of Queden, across from the present Marienwerder, and there, in the year 1233, erected on a hill the fortress called the Island of Holy Mary (Marienwerder). But when the noble and militarily skilled knight from Saxony, the Burggrave of Magdeburg called 'he of the little hand,' had arrived at the fortress of Culm surrounded by many knights and knaves, he moved with the master and the brothers and transferred the aforementioned fortress Marienwerder from the island of Queden to the place where it now stands, in the Pomesanian territory called Reisen, changing its location and not its name."

Peter of Dusburg claimed that the large group of noblemen who contributed to the erection of Marienwerder included dukes Conrad of Masovia, Casimirus of Cuiavia, Henry the Bearded of Wroclaw, and Ladislav Odonic of Wielkopolska.[10] Eventually, the original Marienwerder was identified and, according to both the written sources and archaeologists, it was located at a steep right bank of the Vistula. Therefore, *castrum parvum Quindin* was not located on any island. A similar case is Bischofswerder (Bishop's Island) which was not

---

[9] Translation J. Ogier. Wenta 2007, 56-57 (III, 9).
[10] Wenta 2007, 56-57 (III, 9-10).

located on any island either.[11] Thus, how can we connect the story described by Peter of Dusburg, who reported the events a hundred years after they had happened and combined them with the local tradition, with *castrum parvum Quindin* erected at a steep bank of the Vistula? The solution to this riddle may be hidden in the name of Ostrow-Broze, a locality north of Marienwerder and far from the water. However, Polish Ostrow may be, but does not necessarily have to be an island. Primarily, it signified a stronghold and was usually translated into German as *Werder*. That brings us to the crucial point.

The thirteenth century in Prussia was a period of both coexisting, parallel structures and the spreading of cultural aspects typical of Prussia proper, Germany, and the neighbouring Polish duchies. The domination of one ethnic group did not mean the elimination of others, even in the well-researched territory of northern crusades. The system of group conversions let people retain their former beliefs and convictions. This is even implied in various versions of the names of localities. In Pomesania Christburg (Christ's Castle) was the same as Gotiswalde (God's Forest).[12] In Prussia names such as Engelswalde (Angel's Forest) and Engelsburg (Angel's Castle) do not need to be identified with Christianity.[13] Engelsburg (Copriven or Pokrzywno) may be associated with the Prussian name *rapa*, an angel, i.e., one of the incarnations of Perkunas, the winged Lord of Wind. In nineteenth century Lithuanian legends, God the Father gave His thunderbolt to Perkunas the Angel so that they could collaboratively fight revolts in heaven and prevent evil from spreading on earth.[14] Thus, it is no accident that German Rapotendorf is translated into Polish as Aniolowo.

This picture of the co-existence of pagan and Christian religious values is based on the findings of Anatoly Voluev, an archaeologist who conducted his work in the Alt-Wehlau burial ground, the place where, after the conquest of the King of Bohemia, Přemysl Otakar II, which took place around 1255, Prussians established the town of Wilow in order to prevent a further invasion of the crusaders. Soon after that event, the town's ruler converted to Christianity, becoming a fervent supporter of the Teutonic Order and fighting his pagan tribesmen.

The burial ground existed from the mid-thirteenth till the early seventeenth century. Its most distinctive feature is the co-existence of Christian symbols and

---

[11] Waldemar Heym, Castrum parvum Quidin. Die älteste Burg des Deutschen Ritterordens in Pomesanien, Zeitschrift des Westpreussischen Geschichtsvereins, 70 (1930), 5-68.

[12] Max Hein (ed.), Preussisches Urkundenbuch, Königsberg 1944-1961, IV. nr 403.

[13] Jan Kochanowski (ed.), Codex diplomaticus et commemorationum Masoviae generalis, Warszawa 1919, I. 214-218, nr 217. Maria Rzeczkowska-Sławińska, Zamek w Pokrzywnie, Rocznik Grudziądzki, 8 (1983), 5-31.

[14] Suchocki 1991, 73, 145.

Christian burial rites with non-Christian practices. Until the early sixteenth century, 60 per cent of the deceased had been buried with knives and Teutonic bracteates. Thirteenth century tombs contained equipment such as spear tips, battle knives, and swords. Some of the weapons were ritually bent, singed, or broken. Warriors buried in the fourteenth century were also equipped with weapons, including knights' belts and items ornamented with the symbol of the cross. In the tombs where no warrior equipment was buried, items such as flint and steel, knives, folded razors, flints, and awls could be found. Some of the deceased buried in the fourteenth century had coins put on their chests. Tombs were equipped with grave goods until the seventeenth century. In a tomb of the sixteenth century, a horse bit with rings was found near the warrior's skull. Hollows with the relics of funeral rites, including charcoal, bird bones, and pieces of broken vessels, were dug next to almost 40 per cent of the tombs. The church in Alt-Wehlau was constructed in 1361, which could contribute to the disappearance of the phenomena described above. Those historians are wrong who reject syncretism and assume that this situation resulted from Christianization which lasted for generations.[15] Others assumed that, despite the fact that the Christian faith was confessed, the rituals of the ancestors were still practised. Still, it seems that all forms of the co-existence of both religious systems could occur even though this might not have been realized.

The gradual disappearance of the relics of funeral rites and a longer functioning of elements typical of syncretism with dominant Christianity was a common situation and should be treated as such. However, in the case of the Teutonic State it is a fascinating phenomenon which allows us to demythologize another field researched by academics dealing with the Northern Crusades.[16]

The Religious Situation in Neighbouring Lithuania

It is impossible to understand the religious situation of Prussia without referring to neighbouring Lithuania. The last pagan funerals of the Baltic dukes were those of the Lithuanian Grand Dukes Algirdas (1377) and Kestutis (1382). Most probably, Jogaila abolished making offerings to gods after the Kreva Union of 1385, under which, in return for the Polish crown, he pledged to receive baptism,

---

[15] Anatolij Voluev, Alt-Wehlau - pogańskie cmentarzysko na obszarze chrześcijańskich Prus w świetle badań archeologicznych, in Mirosław Hoffmann und Janusz Sobieraj (eds.), Archeologia ziem pruskich. Nieznane zbiory i materiały archiwalne, Ostróda 15-17 X 1998, Olsztyn 1999, s. 397-400.

[16] Marceli Kosman, Elementy pogańskie w obrzędach pogrzebowych na Litwie po przyjęciu chrześcijaństwa (do XVI w.), Problemy współczesnej tanatologii, 2 (1998), pp. 217-222.

marry the heiress of the Crown, Jadviga Angevin, and join the Grand Duchy to the Crown. Two years later he liquidated the major centres of the pagan cult in the heart of the Grand Duchy. This meant the termination of the co-existence of the dynastic religion, i.e., the former religion of the Balts, with Orthodox Christianity, Catholicism, and Islam. Today we know about the presence of Dominican and Franciscan monks in Lithuania at the beginning of Gediminas' rule. The latter had churches in Vilnius and Navahrudak (1322). Services were held regularly in Vilnius, and the monks always belonged to the dukes' closest circles. At a time of constant conflicts between the Teutonic order and the dukes, personal relationships between the rulers and religious dignitaries went beyond their faith, which was also typical of the manner of conducting war in mediaeval times. For instance, Kestutis had a close relationship with the Commander of Ostroda, Günther von Höhenstein, who was the godfather of his daughter Danuta, who was, in turn, married to Janus, the Duke of Masovia, a Polish region bordering Lithuania and Prussia. The religious system of the Balts met all social expectations; thus, another nobleman to believe in resurrection and the basic rules of Christian faith was the Grand Duke Vytautas Alexander. "Persuaded" by Matthias of Trakai, the Bishop of Vilnius, he became a believer on his deathbed. It may be very enlightening for us to learn that the first part of the Grand Duke's funeral was performed in accordance with the pagan rite; then the coffin was transported to Vilnius, and the corpse was entombed in the cathedral. We can observe conversions from the ancestors' religion to Catholicism among the Grand Duchy elite until as late as the mid-fifteenth century. On his visit to Lithuania in 1547, the Polish king, Sigismund II Augustus, observed that the people in Samogitia still worshipped holy groves, springs of water, and boulders, as well as snakes, and made public and private offerings for them. The Christianization of this land, bordering on Prussia, began only after it was taken over from the Teutonic Order in 1411. The new religion was introduced by force; Jogaila gave orders to cut down the holy groves, quench the holy fires, and expel the pagan priests; however, to little avail. Supported by the Teutonic Order, the adherents of the old religion managed to foment an armed uprising. And on the other hand, Vytautas, the Grand Duke, hindered the activity of overzealous missionaries, expelling them from Lithuania.

Religious Practices and Settlement Changes in the 15th Century

In early fifteenth-century Prussia there were richly equipped body-burning funerals of the representatives of the elite. The coins of the Great Master, Michael Küchmeister, dating from 1414-1416, leave no doubt as to that[17].

---

[17] Stefan Chmielewski, Czy pruski grób ciałopalny z XV wieku. Próba interpretacji

Until the secularization of the Teutonic State in Prussia, parishes included either a settlement (*komornictwo*) or anything from a couple to thirty-five villages. Usually, the people of German and Prussian origin had separate parishes; however, time and colonization contributed to the establishment of combined parishes. Until the late fourteenth century it was common to delegate two persons from each village to attend the Sunday Mass, which reinforced traditional religious practices. It was opposed by the synod of Sambia, called by Bishop Henry Sorborn. According to a document drafted in 1426 by another Sambian Bishop, Michael Junge, offering sacrifices, gathering around burial mounds as well as invoking gods and demons was still common. As has already been observed, the building of mounds and the burning of corpses happened in Prussia until as late as the mid-fifteenth century. Burials combining both religious rites were equally common.[18] Undoubtedly, after the conquest, the Prussians did not provide any single binding system of conduct towards the new rulers or the new religion. The Teutonic Order itself did not conduct any major missionary activity among the Prussians, either. The language barrier also played a significant role in this respect. The German clergy delivered their sermons in German and wanted to have them translated into Prussian, thus needing Prussian-speaking curates. The real breakthrough came with the Reformation.

The result of the conquest and the relocation and defection of people to the neighbouring countries was a decrease in population from approximately 170 thousand at the beginning of the Prussian Crusade to approximately 90 thousand at the end of the thirteenth century. As a consequence of the relocation of Yatvingians to Sambia, the tribal territories of Yatvingians, Galindians, and Sassenians became the Great Wilderness (*die Große Wildnis*). The emigration to Lithuania and Rus led to the establishment of densely built-up Prussian settlements near cities such as Hrodna. Approximately five thousand Yatvingians settled in that region and in Lithuania. By accepting Christianity, a considerable group of noblemen became faithful to the Teutonic Order, even during Prussian uprisings. Assimilation, which ended with adopting the German language and the customs of the German knighthood, affected Matto, Pippin's son, who entered the family of the house of Dytryk von Depenow, whereas Klec, the Prussian, married the daughter of Dytrych Stango, accepting the family's coat of arms and founding the house of Pfeilsdorf-Pilewski. The houses of von Schlubut, Packmohr, Portegal, Wapel, and Perbrandt also had Prussian ancestors.

The fourteenth century witnessed a real renaissance of the Prussian people. In the early fifteenth century 140 thousand Prussians constituted about 50 per cent of the total population of Prussia. There was also a strong group of so-called

---

znaleziska z 1703 roku, Rocznik Olsztyński, 5 (1963), 295-320.
[18] Okulicz-Kozaryn 1997, 459-477. Marceli Kosman, Polska-Litwa. Z odległej i bliższej przeszłości, Poznań 2003, I.117-127, 162.

freeholders, i.e., owners of estates, who had to serve in the army. In the early fourteenth century, Prussian freeholders received their estate under the Chelmno Law or the less advantageous Magdeburg Law (*ius Magdeburgense simplex*). The Prussian law was codified and the court (*Wayda*) was established. Despite the fact that Prussian clergymen constituted a small group, some of them rose to high Church ranks. The Prussian population in towns and cities made up less than ten per cent. At the same time, the German population constituted as much as fifty per cent of city dwellers. The Great Wilderness was colonized with the migration of Poles and Lithuanians to the south and east of Prussia, respectively. The wars at the end of the fifteenth and the beginning of the sixteenth century resulted in Prussians losing their privileged position and becoming indigent, which triggered assimilation. Thus, the Prussian freeholders became peasants.

After the Thirteen Years' War (1454-1466) more and more Prussians would pass themselves off as Poles or Germans. Under the rule of the last Grand Master of the Teutonic Order and the first Duke of Prussia there were few Prussian-speaking priests, whereas the absence of interpreters (*tolks*) in the territories inhabited by Prussians implies that they must have been Germanized. In the mid-seventeenth century their language became extinct. We do not know why people stopped speaking Prussian. We realize that the decline of the language occurred rapidly after 1570, and the language became extinct within one generation. After 1545 the sources do not mention any services in Prussian. We also know there was no pressure to use German; however, there were cases in Warmia[19] where rulers were forbidden to communicate with their subjects in Prussian. References to holiness appearing in German names of localities imply that adopting the language was not tantamount to renouncing traditional behaviour. Thus, German Heiligenthal, Heilsberg, Heiligenbeil or Polish Schwente (*silova sancta ultra Prigoram*) should be associated with the pre-Christian sacred.[20]

In conclusion, I would like to take you to the year 1520 when Sambia was threatened with an invasion of Polish worship practices. Having obtained the Grand Master's consent, a Prussian priest, Valtin Supplit, made an offering of a black bull to the Lord of Wind. The Church investigation confirmed the common

---

[19] Karol Górski, Zanik dawnych Prusów, Zapiski Historyczne, 47 (1982), 597-604; Grzegorz Białuński, Drogi zaniku Bałtów (XII-XVII), Studia Angeburgica, 7 (2002), 7-20.

[20] Waldemar Rozynkowski, Hagiotoponimia w państwie zakonu krzyżackiego w Prusach. Zarys problematyki, In Urszula Borkowska and Czesław Deptuła, Peregrinatio ad veritatem. Studia ofiarowane profesor Aleksandrze Witkowskiej OSU z okazji 40lecia pracy naukowej, Lublin 2004, pp. 459-467. Carl Peter Wölky und Johann Martin Saage (ed.), Codex diplomaticus Warmiensis, Mainz 1860, II. nr 222; Carl Peter Wölky (ed.) Preussisches Urkundenbuch, Königsberg 1889, I/2, s. 310; August Seraphim (ed.), Preussisches Urkundenbuch, Königsberg 1909, II. nr 704.

conviction about the priest's efficacy. The same priest is mentioned in the sources again in 1530-1531 when he prevented the spread of the plague affecting fishermen in the Courland Lagoon.[21]

---

[21] Suchocki 1991, 134, 168-169.

Bibliography

I. Sources

Max Hein (ed.), Preussisches Urkundenbuch, Königsberg 1944-1961, IV.

Jan Kochanowski (ed.), Codex diplomaticus et commemorationum Masoviae generalis, Warszawa 1919, I.

August Seraphim (ed.), Preussisches Urkundenbuch, Königsberg 1909, II.

Max Töppen (ed.), Petri de Dusburg Chronicon terrae Prussiae, Scriptores rerum Prussicarum, Leipzig 1861, I. 3-269.

Jarosław Wenta, Sławomir Wyszomirski (ed.), Piotr z Dusburga, Kronika ziemi pruskiej, Monumenta Poloniae Historica, n.s., t.XIII, Kraków 2007.

Carl Peter Wölky und Johann Martin Saage (ed.), Codex diplomaticus Warmiensis, Mainz 1860, II.

Carl Peter Wölky (ed.) Preussisches Urkundenbuch, Königsberg 1889, I/2.

II. Secondary Literature

Joseph Bender, "Zur altpreussischen Mythologie und Sittengeschichte," Altpreussische Monatsschrift, 2 (1865), 577-603, 694-717, 4 (1867), 1-27, 97-135.

Grzegorz Białuński, "Drogi zaniku Bałtów (XII-XVII)," Studia Angeburgica, 7 (2002), 7-20.

Grzegorz Białuński, "Las w wierzeniach Prusów i Jaświngów," Feste Boyen, 1 (2002), 31-42.

Haralds Biezais, Die Religionsquellen der baltischen Völker und die Ergebnisse der bisherigen Forschungen, Uppsala 1954, s.65-128.

Grasilda Blaziene, Die baltischen Ortsnamen in Samland, Stuttgart 2000.

Aleksander Brückner, Starożytna Litwa, Ludy i bogi: szkice historyczne, Olsztyn 1979.

Aleksander Brückner, "Die Galindensage," Archiv für slawische Philologie, 21 (1899), 22-27.

Franciszek Bujak, Dwa bóstwa prusko-litewskie "Kurche" i "Okkopirnus", Lwów 1924.

Stefan Chmielewski, „Czy pruski grób ciałopalny z XV wieku. Próba interpretacji znaleziska z 1703 roku," Rocznik Olsztyński, 5 (1963), 295-320.

Hans Crome, "Die Religion der alten Preußen," Alt Preußen, 4 (1939), 50-54.

Marian Dygo, "O kulcie maryjnym w Prusach krzyżackich w XIV-XV w.," Zapiski historyczne, 52 (87), 237-268.

Wilhelm Friederici, "Über die Lage Romow's oder Romowe's des Oberpriestersitz im heidnischen Preussen," Altpreussische Monatsschrift, 13 (1876), 227-253.

Georg Gerulis, Die altpreußischen Ortsnamen, Berlin und Leipzig 1922.

Eduard Gigas, "Der sogenante Potrimpos zu Christburg, der sogenante Bartel und die Gustabalde zu Bartenstein," Zeitschrift des historischen Vereins für den Regierungsbezirk Marienwerder, 2 (1877), 43-54.

Karol Górski, "Zanik dawnych Prusów," Zapiski Historyczne, 47 (1982), 597-604.

Waldemar Heym, "Castrum parvum Quidin. Die älteste Burg des Deutschen Ritterordens in Pomesanien," Zeitschrift des Westpreussischen Geschichts-vereins, 70 (1930), 5-68.

Marceli Kosman, "Elementy pogańskie w obrzędach pogrzebowych na Litwie po przyjęciu chrześcijaństwa (do XVI w.)," Problemy współczesnej tanatologii, 2 (1998), 217-222.

Marceli Kosman, Polska-Litwa. Z odległej i bliższej przeszłości, Poznań 2003.

Jerzy Marek Łapo, "Czcij praojca swego: o kulcie przodków w kręgu zachodniobałtyjskim na początku I tys. n.e.," Popiół i kość: Funeralia lednickie, 4 (2002), 321-337.

Łucja Okulicz-Kozaryn, Dzieje Prusów, Wrocław 1997.

Jerzy Suchocki, Mtologia bałtyjska, Warszawa 1991.

Jan Powierski, Bogini Kurko i niektóre aspekty społeczno-gospodarcze wierzeń pruskich.

Jan Powierski, "Chronologia początków Malborka," Zapiski Historyczne, 44 (1979), 181-207.

Rozalia Przybytek, Ortsnamen baltischer Herkunft im südlichen Teil Ostpreußen, Stuttgart 1993.

Adolf Rogge, "Der preussische Landberg, das älteste Romowe," Altpreussische Monatsschrift, 14 (1877), 585-592.

Waldemar Rozynkowski, Hagiotoponimia w państwie zakonu krzyżackiego w Prusach. Zarys problematyki, in Urszula Borkowska and Czesław Deptuła (eds.), Peregrinatio ad veritatem. Studia ofiarowane profesor Aleksandrze Witkowskiej OSU z okazji 40lecia pracy naukowej, Lublin 2004, 459-467.

Maria Rzeczkowska-Sławińska, "Zamek w Pokrzywnie," Rocznik Grudziądzki, 8 (1983), 5-31.

Bernhard Schmid, "Die Gründung der Marienburg," Altpreußische Forschungen, 6 (1929), 191-200.

Jerzy Suchocki, Mtologia bałtyjska, Warszawa 1991.

Anatolij Voluev, "Alt-Wehlau - pogańskie cmentarzysko na obszarze chrześcijańskich Prus w świetle badań archeologicznych," in Mirosław Hoffmann and Janusz Sobieraj, Archeologia ziem pruskich. Nieznane zbiory i materiały archiwalne, Ostróda 15-17 X 1998, Olsztyn 1999, 397-400.

Jarosław Wenta, Kronika Piotra z Dusburga, Toruń 2003.

Benno Winckler, "Romowe in Warmien," Zeitschrift für die Geschichte und Altertumskunde Ermlands, 3 (1866), 521-526.

Prof. dr. hab. Jarosław Wenta
Uniwersytet Mikołaja Kopernika
Centrum Mediewistyczne Wydziału Nauk Historycznych
Ul. Szosa Chełmińska 83 a
PL - 87-100 Toruń, Poland
E-Mail: jwe@his.uni.torun.pl

Dante and the Island of Purgatory

"Chi vive della fede non cerca di scrutarla con l'argomentazione e di concepirla con la ragione; preferisce prestar fede ai misteri celesti piuttosto che affaticarsi vanamente, mettendo in disparte la fede, a comprendere ciò che non può essere compreso".[1]

Probably the most evocative image of the island on which the mount of Purgatory rises, as expressed by Dante in the second *canto* of *The Divine Comedy*, is that which appears to Ulysses just before the shipwreck. The Homeric hero, driven to exceed limits that are humanly possible by his thirst for knowledge, will never return to Ithaca in the Dantesque version, but turn the prow of his ship to beyond the pillars of Hercules, sail for five months in an unknown sea and finally sink in a vortex of wind and water before the very mount which had raised hope in the sailors when they first saw it:

> *"quando n'apparve una montagna, bruna*
> *per la distanza, e parsemi alta tanto*
> *quanto veduta non avea alcuna.*
> *Noi ci allegrammo, e tosto tornò in pianto;*
> *chè della nova terra un turbo nacque,*
> *e percosse del legno il primo canto.*
> *Tre volte il fe' girar con tutte l'acque:*
> *alla quarta levar la poppa in suso*
> *e la prora ire in giù. Com'altrui piacque,*
> *infin che 'l mar fu sopra noi richiuso".*[2] (Inferno, XXVI, 133-142)

---

[1] Lanfranco da Pavia (1010-1089) P.L. 150, De corp et sang. Dom.,17. "Who lives in faith does not try to seek it out with reason and to conceive it with rationality; he would prefer to believe in celestial mysteries rather than exhausting himself hopelessly, putting aside faith, trying to understand what cannot be comprehended" (my translation).

[2] Quotations from Dante Alighieri, *La Divina Commedia*, N. Sapegno (ed.), La nuova Italia Editrice, Firenze, Inferno 1974[7] Purgatorio 1975[8], Paradiso 1976[8].
"When a lone mountain loomed ahead, dark / In the dim distance, and it looked to me / The highest peak that I had ever seen. / We leaped for joy it quickly turned to grief, / For from the new land a whirlwind surging up / Struck the foredeck of our ship head on. / Three times it spun us round in swirling waters; / The fourth round it raised the stern straight up / And plunged the prow down deep, as Another

This apocalyptic image concludes the canto dedicated to Ulysses in hell, the real anti-hero of the *miles christianus*, his sin being irremediably compared to that supreme sin committed by Lucifer; that of having refused to confide in God and, instead, depending on his own strength.

The identification of an intermediary kingdom between heaven and hell already existed amongst the pagans[3], which led to the conception of lake Acherusiade, in which sinners temporarily stayed to purify themselves:

"Dans *Fedone*, Platon parle des traditions qui couraient de son temps sur le séjour des morts. La triple division que le christianisme a faite de l'autre monde s'y trouve déjà marquée: le lac Achérusiade, où les coupables sont temporairement purifiés, c'est le purgatoire; le Tartare, d'où ils ne sortent jamais, c'est l'enfer; enfin ces pures demeures au-dessus de la terre, qui ont elles-mêmes leur degré de beauté selon le degré de vertu de ceux qui les habitent, c'est le paradis. Seulement Platon ajoute prudemment: 'Il n'est pas facile de les décrire.' Peut-être est-ce le mot qui a piqué l'émulation de Dante."[4]

The Scriptures do not explicitly speak of Purgatory, but some passages state the possibility of expiating the sins of the dead. In the writings of the church fathers we can find the first traces of the belief in the purification of the soul, a conviction confirmed by Saint Gregory the Great in the sixth century.[5]

Le Goff makes reference to a development of this tradition in two phases:[6] one being a theological-spiritual development, bound to the Cistercian movement in Paris between 1170-1180, whereas the other is connected to the literature of visions between 1180-1215. The Council of Lyon (1274), then of Florence

---

pleased, / Until the sea once more closed over us." The Divine Comedy, Translated by James Finn Cotter, Web Edition by Charles Franco, http://www.italianstudies.org/comedy/index.htm.

[3] See Claudia di Fonzo, *La leggenda del "Purgatorio di S. Patrizio" nella tradizione di commento trecentesca*, lecture held in the Department of Italian Literature of "La Sapienza" in Rome the 10th June 1997, published in *Dante e il locus inferni. Creazione letteraria e tradizione interpretativa* a cura di S. Foà e S. Gentili, «Studi (e testi) italiani» 4 (1999), pp. 53-72, note 2.

[4] See C. Labitte, La Divine Comédie avant Dante, in Revue des Deux Mondes, IV Série, 31 (1842), pp. 704-742; see also *Oeuvres de Dante Alighieri. La Divine Comédie. La Vie Nouvelle*, nouvelles éditions revues, corrigées et annotées par les traducteurs et d'une étude sur la Divine Comédie Paris, Charpentier, 1872, pp. 85-150, p. 91.

[5] See note 8 in this article.

[6] See J. Le Goff, *La nascita del Purgatorio*, Torino, Einaudi, 1982 , pp. 203-204 (original text *La naissance du Purgatoire*, Paris, édition Gallimard, 1981); see also *L'immaginario medievale*, Roma-Bari, Laterza, 1988. (original text L'immaginaire médiéval, Paris, Éditions Gallimard, 1985).

(1439), and finally that of Trent (1553) indicate that the place of expiation is in Purgatory.[7]

It would be interesting, at this point, to ask why Dante chose this particular setting for his Purgatory, a mountain, that is to say, on a mountain-island in the middle of nowhere, placed in the middle of a hemisphere of water, poles apart from Jerusalem, when, according to common theological opinion, Purgatory is underground and part of hell, as asserted in Book IV of Pope Gregory's Dialogue.[8]

In the Dantesque notion, however, Lucifer's fall to earth, which happened immediately after the creation, in a certain way gave rise to the abyss of Hell, with the displaced material, having been pushed away and upwards, producing a high mountain which rose alone in the southern hemisphere.

The souls of the dead who were graced by God are grouped together at Ostia at the mouth of the river Tiber by a guiding angel before reaching the mount. This angel then leads them to the shore of Anti-purgatory after having supervised their boarding onto fast and light sailing vessels.

Dante divides the mount into three parts: Anti-purgatory, Purgatory and earthly Paradise. Anti-purgatory consists of the shore of the island and of two cliff edges, where we find those who repented when they died.

Purgatory consists of seven concentric circles, but not spirals, which host sinners according to the order of the deadly sins: pride, envy, anger, sloth, avarice and extravagance, gluttony, and lust.

In this kingdom the division of sins refers to the principle of good, because love can sin for *malo obbietto*, when it has a leaning towards being bad, or for *bono obbietto*, when it has a leaning towards the highest good with *poco di vigore* or with *troppo di vigore*; if one turns towards evil, one gives way to the sins of pride, envy, and anger; if one turns to God with little strength, one gives way to sloth; if one turns towards earthly good with too much vigour, one gives way to avarice, gluttony, and lust. Love, however, cannot be a source of sin if one reaches towards God in the right measure. The fundamental medieval concept of *maze* is clearly demonstrated here, that is, of the right balance between impulsiveness and willpower, which distinguishes the core of many chivalric epics.

---

[7] See Roberto Beretta's interview with Franco Cardini, in *Il Timone*, 14, July/August 2001.

[8] See Claudia di Fonzo, as in note 3. In the fourth book of *Dialoghi,* attributed to Gregory Magnus, is told the story of a monk who died after committing a horrible sin against poverty, without being reconciled with the Church. After thirty days, during which a daily mass for the monk's soul was celebrated, a brother appeared claiming his freedom from Purgatory's punishments (see *Dialoghi* IV, 55).

58

As his source, Dante makes explicit reference to journeys made to the afterlife by Saint Paul and by Aeneas, according to tradition[9]. However, in the *Comedy* many varied accounts of elaborate visions from Christianity come together to provide tangible images of the hereafter and to make it more comprehensible. Yet Dante's great poem surpasses the representations produced to affect people of little or no education, such as *Saint Paul's Vision*; the *Apocalypse of Saint John* and *The Vision of Friar Alberico of Montecassino*; the *Navigatio Sancti Brendani* to the lost island, which medieval maps mark in the Atlantic near the Canary Islands; *The Legend of Saint Patrick's Well* and that of *Owen's Well*, where after climbing down into it, one enters earthly Paradise by a very narrow bridge.[10]

The so-called *Saint Patrick's Well* was a very deep cave found on the isle of Derg, in the northwest of Ireland. According to the Irish tradition surrounding the cult of the island's patron saint, Christ revealed the deep cave to Patrick in order to help him convince the unbelieving Irish of what great suffering in the afterlife awaited those who were tainted with sin in this life; in fact it was said that in this cave there was a bottomless well, from which opened the doors to Purgatory, and that whosoever stayed there one whole day and whole night had his sins pardoned.[11]

The legend of the Irish Knight called Owen and his visit to Purgatory, to which he gains access through the same gap that had at the time been shown to Saint Patrick, and after the visit to which it was his task to recount to the living all he saw and heard, became famous in about the twelfth century. The text was

---

[9] *Io non Enea, io non Paulo sono*, I am not Aeneas, nor am I Paul! Inferno, II, 32, with reference to *Eneide*, book IV and to II Corinthians, 12, 2-4. Quotations as in note 2.

[10] See *Il Timone*, above, footnote 7, and Claudia di Fonzo, above, footnote 3, in particular notes 9-17. *Saint Paul's vision* is an Apocrypha of the New Testament. Originally written in Greek around the middle of the third century, it has come down to us in many versions and it had much success during the Middle Ages; it narrates Paul's journey from heaven to hell. The *Visione di Alberico*, which narrates Alberico's journey in the afterlife when he was ten years old, is in the *Codice Cassinese 257*, of which a copy is preserved in the Alexandrine Library of Rome; see the version edited from father Mauro Inguanez Osb in "Miscellanea Cassinese" I (1932), p. 83-103; *Saint Brendan's Voyage* is narrated in *Navigatio sancti Brendani* written by an anonymous Irish author of the tenth century and describes the saint's marvelous apostolic journeys.

[11] See Claudia di Fonzo, note 16, with reference to St Patrick's Purgatory, *Two versions of Owayne Miles and the Vision of William of Stranton together with the long Text of the Tractatus De Purgatorio Sancti Patricii*, edited by Robert Easting, The Early English Text Society, Oxford University Press, 1991, pp. 121-154 (text) and pp. 236-254 (comment).

handed down in various languages, but one of the most important versions is that of Marie de France, which enabled it to be circulated in Europe. In fact after Marie's translation other popular versions appeared, especially in French and English.[12]

This legend, a bestseller in the medieval period, underwent numerous adaptations.

In Italy, for example, it can be found in Jacobus de Voragine's *Legenda aurea* (1260-1275), as well as in a Venetian code from the fifteenth century, in which the protagonist Owen, by now very well-known in Venice, is called Alvise.[13]

It could actually have been this text which influenced the *Comedy*, because the ancient commentators of Dante's work, for example Alberico da Rosciate and Benvenuto da Imola, name the following as the sources for Purgatory: the first letter of the Corinthians (3,13-15); the psalm *Iubilate deo ver*, and in particular the verse *Trensivimus per ignem et aquam et deduxisti nos in refrigerium*; and the legend of Saint Patrick.[14]

The iconography, which spread throughout Italy in relation to the legend, shows Purgatory to be underground, but at the same time on a promontory. This same image can be found on a fresco in the choir of the monastery of Saint Francis at Borgo Nuovo in Todi. This fresco, believed to be from 1346 and discovered in 1975, has been attributed to Iacopo di Mino of Pellicciaio. It covers a whole wall and shows the passage of souls from Purgatory to Paradise with the Virgin Mary and Saint Philip Benizi acting as their go-betweens. Purgatory is pictured as a mountain with seven caves, on the top of which there is a well, where we can see Saint Patrick and Nicolaus the Knight.[15]

---

[12] See Claudia di Fonzo, above, footnote 3, with reference to, among other titles, T. A. Jenkins, *The Espurgatoire Saint Patriz of Marie de France, with a Text of the Latin Original* in "The Decennial Publications of the University of Chicago", first series, VII (1903), pp. 235-327; see also Jacques le Goff, as well as note 6, p. 224.

[13] See Claudia di Fonzo, above, footnote 3, notes 39-47, with reference to Jacobi da Varagine, *Legenda Aurea. Vulgo historia lombardica dicta*, recensuit Dr. Th. Graesse, Osnabrück, Otto Zeller, 1965 and Giusto Grion, *Il Pozzo di S. Patrizio*, in "Il Propugnatore", anno 3.° dispensa Iᵃ (1870), pp. 67-149; Grion's version is edited from a Venetian code of the fifteenth century.

[14] See Claudia di Fonzo, above, footnote 3, notes 50-60, with reference to M. Petoletti, «*Ad utilitatem volentium studere in ipsa Comedia*»: the Dantesque comment of Alberico da Rosciate, in «Italia Medioevale e Umanistica» 38 (1995), pp. 141-216.

[15] See Claudia di Fonzo, above, footnote 3, note 84, with reference to Marcello Castrichini (ed.) "Dal Purgatorio di S. Patrizio alla città celeste, a proposito di un affresco del 1346 ritrovato a Todi", EDIART, 1985.

A later representation can be found in the fresco of the church of St. Maria in Piano di Loreto Aprutino. The bridge, which is actually called the "hair" because it becomes narrower and narrower, clearly recalls the bridge of St. Patrick's legend and is positioned immediately before Paradise. Beneath this bridge is the river of *amaritudo*, bitterness, into which all those souls who have not managed to cross the bridge fall.[16]

Dante, however, managed a very real revolution in respect to the preceding sources. He takes Purgatory away from the underground, he distinguishes it from Hell, and he creates a "stairway" which allows the sinners to get onto the road to Paradise, after death, atoning for their sins by taking a narrow and difficult path that they were not able to follow during their lifetime.[17]

A last reference to the legend can be found in the long and narrow tunnel Dante passes through with Virgil to reach Purgatory from Hell, which he does on land. The poet's emotions on seeing the light of the stars again is famous, anticipating the bright light of Paradise once he has surfaced on the beach of Paradise:

> *"Dolce color d'oriantal zaffiro,*
> *Che s'accoglieva nal sereno aspetto*
> *Del mezzo, puro insino al primo giro,*
> *alli occhi miei ricominciò diletto,*
> *tosto ch'io usci' fuor dell'aura morta*
> *che m'avea contristati li occhi e 'l petto"*
> Purgatorio, I, 13-18[18]

It is also possible that his idea was inspired by the legend of Saint Brendan's voyage, which tells of how he saw a tall, crystal tower in the sea and an island of

---

[16] See E. Cerulli, *Il "Libro della Scala" e la questione delle fonti arabo-spagnole della Divina Commedia*, Città del Vaticano 1949; Claudia di Fonzo, above, footnote 3, note 84, with reference to S. Dell'Orso, *Considerazioni intorno agli affreschi della Chiesa di S. Maria in Piano a Loreto Aprutino*, in Bollettino d'Arte, LXXIII (1988), n. 49, serie VI, pp. 63-82; R. Torlontano, *Abruzzo*, in Pittura murale in Italia. Dal tardo duecento ai primi del quattrocento, Bolis Edizioni, 1995, pp. 177-179.

[17] See Claudia di Fonzo, above, footnote 3; for possible influences of Islamic tradition on the new concept of Dante see also E. Cerulli, *Nuove ricerche sul "Libro della Scala" e la conoscenza dell'Islam in Occidente*, Città del Vaticano 1972 and G. Berti, *I mondi ultraterreni*, Mondatori, Milano 1998.

[18] Soft coloring of oriental sapphire, / Collecting in the calm face of the sky, / Clear right up to the edge of the horizon, / Brought back delight again into my eyes / As soon as I stepped out from the dead air / Which overburdened both my sight and breast: quotations as in footnote 2.

fire. However, what is characteristic to Dante alone is the idea of a high mountain on the island of "the halfway world," the most human of the three afterlife kingdoms, that of hope in which atonement and purification lead to blessedness.

It now becomes important to return to the opening image and to Ulysses' voyage towards the Mount of Purgatory. The voyage of this proud hero towards an unknown island, a voyage which recalls the ancient navigations of the Mediterranean, is in fact contrasted with the voyage of atonement of the *miles christianus* who trusts in God, which condenses classical inheritance with Christian spirituality.

Purgatory, which rises all alone on the other side of the world, could not be anything but a mountain on an island, because it contains the pain of atonement but at the same time the hope of forgiveness; because the island is a place of refuge, but more often a place of waiting; it is safety for the shipwrecked, but also a place of limitation because one is kept far from the rest of the world. It is a world of transition, of sufferance, and of patient transformation. The mountain rises out of the water and climbs high into the sky, taking with it the principle of the transitory.

The cleansed soul, positioned between eternal sorrow and the happiness of divine contemplation, still preserves the memory of sin but, at the same time, aspires to the grace of God. It confides in the memory of the living, because the memory of the living and their pity mirrors divine mercy, ready to welcome prayers and acts of intercession, confirmed by the terror of being forgotten by God that was so widespread in medieval times. The meaning of memory in the Middle Ages is dated back to Saint Augustine's teaching about trinity, in which *memoria, intelligentia,* and *voluntas* are the soul's attributes that come from God, and *memoria* is intrinsically linked to the cult of the dead. The memory of the living leads to the constant presence of the dead in God's memory, so remembering becomes an act of care and worry about the satisfaction of the dead in the afterlife. Mentioning the deceased during the liturgic act leads to God's remembering the deceased as well.[19] In this way Manfredi's words towards Dante, at the end of their encounter, can be understood.[20]

---

[19] See F. Ohly, *Bemerkungen eines Philologen zur Memoria* in Wollasch, Joachim u.a. (Hrsg.): *Memoria. Der geschichtliche Zeugniswert des liturgischen Gedenkens im Mittelalter*, München 1984, pp. 9-68, 28, 34, and 53.

[20] Manfred of Altavilla, natural son of Frederick II, king of Sicily from 1258, was excommunicated by Pope Innocent IV and died fighting against Carl of Angiò in Benevento in 1266. His daughter Constance, mentioned in the chant, was Peter of Aragon's bride.

"Vedi oggimai se tu mi puoi far lieto,
rivelando alla mia buona Costanza
come m'hai visto, e anche esto divieto;
ché qui per quei di là molto s'avanza". (Purgatorio, III, 142-145)[21]

Once again then, Dante's work displays the culmination of the Latin and popular medieval culture as a whole, giving voice to that knowledge profoundly linked to images, in which the "blank spaces" of communication and information are filled by a fanciful creativity, a knowledge which, as asserted by Goethe and Carlyle, was to remain dumb for ten long centuries afterwards.[22]

---

[21] "You see by now how you can make me happy / By letting my kindhearted Constance know / How you have seen me, and this interdict, / For those beyond there much advance us here": quotations as in footnote 2.

[22] See F. Marroni, *The Shadow of Dante: Elizabeth Gaskell and The Divine Comedy*, in the Gaskell Society Journal 10 (1996) pp. 1-13, with reference to *The Works of Thomas Carlyle*, ed. D. H. Traill, London: Chapman and Hall, 30 voll., 1895-9, vol. V, p. 101. Concerning the relationship between Dante and Goethe see also W. Hird, *Goethe und Dante*, in: Deutsches Dante-Jahrbuch, Herausgegeben im Auftrag der deutschen Dante-Gesellschaft e.V. von Marcella Roddewig, Boehlau Verlag Koln Wien 1994, Band 68/69, pp. 31-80.

Bibliography

I. Sources

Dante Alighieri, *La Divina Commedia*, a cura di N. Sapegno, La nuova Italia Editrice, Firenze, Inferno 1974[7] Purgatorio 1975[8], Paradiso 1976[8]. The Divine Comedy, Translated by James Finn Cotter, Web Edition by Charles Franco, http://www.italianstudies.org/comedy/index.htm

Gregorio Magno (san), *Storie di santi e di diavoli. Dialoghi*, 4 vols, collana scrittori greci e romani, Mondatori, Milano 2006.

Il Pozzo di S. Patrizio, edito da G. Grion, in "Il Propugnatore", anno 3.° dispensa I[a] (1870), pp. 67-149.

Jacobi da Varagine, *Legenda Aurea. Vulgo historia lombardica dicta*, recensuit Dr. Th. Graesse, Otto Zeller, Osnabrück 1965

Lanfranco da Pavia (1010-1089) P.L. 150, *De corp et sang. Dom.*,17.

St Patrick's Purgatory, Two versions of Owayne Miles and the Vision of William of Stranton together with the long Text of the Tractatus De Purgatorio Sancti Patricii, edited by Robert Easting, The Early English Text Society, Oxford University Press, 1991, pp. 121-154 (text) and pp. 236-254 (comment).

Virgil's Aeneid trans by J. Dryden, ed. by Frederick M. Keener, Penguin Classics, London 1997.

II. Secondary Literature

Berti, G., *I mondi ultraterreni*, Mondatori, Milano 1998.

Castrichini, M. (a cura di), *Dal Purgatorio di S. Patrizio alla città celeste, a proposito di un affresco del 1346 ritrovato a Todi*, EDIART, 1985.

Cerulli, E., *Nuove ricerche sul "Libro della Scala" e la conoscenza dell'Islam in Occidente*, Città del Vaticano 1972.

Cerulli,_E., *Il "Libro della Scala" e la questione delle fonti arabo-spagnole della Divina Commedia*, Città del Vaticano 1949.

Dell'Orso, S., *Considerazioni intorno agli affreschi della Chiesa di S. Maria in Piano a Loreto Aprutino*, in: Bollettino d'Arte, LXXIII (1988), n. 49, serie VI, pp. 63-82.

Fonzo, di, C., *La leggenda del "Purgatorio di S. Patrizio" nella tradizione di commento trecentesca*, Comunicazione tenuta presso il Dipartimento di Italianistica della Sapienza di Roma il 10 giugno 1997, pubblicata in *Dante e il locus inferni. Creazione letteraria e tradizione interpretativa* a cura di S. Foà e S. Gentili, «Studi (e testi) italiani» 4 (1999), pp. 53-72.

Hird, W., *Goethe und Dante*, in: Deutsches Dante-Jahrbuch, Herausgegeben im Auftrag der deutschen Dante-Gesellschaft e.V. von Marcella Roddewig, Böhlau Verlag Köln Wien 1994, Band 68/69, pp. 31-80.

Intervista di Roberto Beretta a Franco Cardini, mensile cattolico Il Timone, n. 14, Luglio/Agosto 2001.

Jenkins, T. A., *The Espurgatoire Saint Patriz of Marie de France, with a Text of the Latin Original,* in "The Decennial Publications of the University of Chicago", 1st series, VII (1903), pp. 235-327.

Labitte, C., La Divine Comédie avant Dante, in Revue des Deux Mondes, IV Série, 31 (1842), pp. 704-742; poi in *Oeuvres de Dante Alighieri. La Divine Comédie. La Vie Nouvelle*, nouvelles éditions revues, corrigées et annotées par les traducteurs et d'une étude sur la Divine Comédie Paris, Charpentier, 1872, pp. 85-150

Le Goff, J., *La nascita del Purgatorio*, Torino, Einaudi, 1982 , pp. 203-204 (testo orig. *La naissance du Purgatoire*, Paris, édition Gallimard, 1981);

Le Goff, J., *L'immaginario medievale*, Roma-Bari, Laterza, 1988. (Testo orig. L'immaginaire médiéval, Paris, Éditions Gallimard, 1985).

Marroni, F., *The Shadow of Dante: Elizabeth Gaskell and The Divine Comedy*, in the Gaskell Society Journal 10 (1996) pp. 1-13.

Mazzadi, P., *Autorreflexionen zur Rezeption: Prolog und Exkurse in Gottfrieds 'Tristan'*, Trieste, Edizioni Parnaso 2000.

Ohly, F., *Bemerkungen eines Philologen zur Memoria*, in: Wollasch, Joachim u.a. (Hrsg.): *Memoria. Der geschichtliche Zeugniswert des liturgischen Gedenkens im Mittelalter*, München 1984, S. 9-68.

Petoletti, M., *«Ad utilitatem volentium studere in ipsa Comedia»*: il commento dantesco di Alberico da Rosciate, «Italia Medioevale e Umanistica» 38 (1995), pp. 141-216.

Torlontano, R., *Abruzzo*, in: Pittura murale in Italia. Dal tardo duecento ai primi del quattrocento, Bolis Edizioni, 1995, pp. 177-179.
Traill, D.H., (ed.) *The Works of Thomas Carlyle*, 30 vols, Chapman and Hall, London 1895-9.

Dr. Patrizia Mazzadi
Università degli Studi di Urbino
Via Leopardi 6
I – 36100 Vicenza
patrizia.mazzadi@uniurb.it

The Island of Cyprus in Travel Literature of the Fourteenth Century

I. Introduction

While the admission of Turkey to the European Union is being debated, the Cyprus question stands in the foreground, and the island has become a rather hot topic. Located in the eastern Mediterranean near the coasts of Turkey and Syria, Cyprus occupies a strategic and logistically important position between Asia and Africa, yet it belongs to Europe. Its position has repeatedly led to political disputes and tensions during the course of its history. In the fourteenth century, which will be the main focus of this study, Cyprus was a kingdom under the rule of the French dynasty of the Lusignans, who had ruled the island since 1192.[1]

One of the most prominent personages of the fourteenth century, Francesco Petrarch, wrote this about the island kingdom:

Beyond the coast of Cilicia lies Cyprus, a land known for its laziness and lusts. It has been rightly said that it is sacred to Venus and even today, it shows more devotion to Venus than to Mars or Athena. It has seldom, if ever, produced a famous man, since the robust seeds of virtue rarely flourish in the soft fields of lust. The heat of earth and air are a sign of the desires of the inhabitants. While regions far from the sun enjoy pleasant, moderate temperatures, this area glows in almost unnatural and unbearable heat. The condition of the people is almost one with the elements. I would not like to stay here very long. For this is certainly no habitation for a militant and manly character: the Gallic arrogance, the Syrian softness, Greek flattery and deception are united on one island. The best and most valuable thing they have is this: having come from elsewhere and with other customs, St. Hilarion is buried here.[2]

---

[1] Cf. Maier, 1982, p. 93-128. The Lusignans ruled Cyprus until 1489. The last ruler, Caterina Cornaro, the widow of King James II, abdicated in that year in favor of Venice, Feldbauer/Morrisey, p. 67, 71; Kretschmayr II, p. 390f. After the reign of the Venetians, Cyprus was conquered by the Ottoman Empire in 1571, ibid. p. 73 and von Reden, p. 255-260. Concerning the short rule of the Templars on Cyprus before the Lusignans, see also Tzermias, p. 9.

[2] Francesco Petrarca: Itinerarium ad Sepulchrum Domini nostri Iesu Christi, (Itin. 52): *Ante Ciliciae frontem Cyprus est, terra nulla re alia quam inertia ac deliciis nota, quam merito Veneri sacram dixere et nunc quoque Veneri magis quam Marti seu Palladi sacra est. Raro ibi seu nunquam vir aliquis clarus fuit, neque enim in molli agro voluptatis virtutum rigida semina coalescunt. Libidinem incolarum terrae coelique fervor indicat. Cum enim regiones tractu maximo soli viciniores*

Petrarch attaches to the island less than flattering characteristics such as indolence, idle luxury, heat, and lust, and draws a very negative picture of its inhabitants, generalizing national characteristics or prejudices, at least in his "Itinerarium", written in 1358. Petrarch names the fourth century St. Hilarion,[3] who came from Gaza (in today's Palestine), as its only positive feature. Of course, it is necessary to take into account Petrarch's readership and his intention, that is, to describe the virtuous man whose *virtus* overcomes and outlasts any adversity. Nevertheless, we can compare the picture of the island painted by Petrarch, who never visited Cyprus, with those of his contemporaries who did have a chance to spend time on the island.

Therefore, the following study will center on travel descriptions of the fourteenth century or voices from the Middle Ages and the presentation of their pictures of Cyprus, their positive and negative impressions, and their portrayals of the mentality of the people. These may contradict or reinforce those of Petrarch.

Since the travel descriptions of Ludolf von Sudheim and Jacopo da Verona provide especially comprehensive portrayals, they will be the focus of the following discussion. The problems common to the travel and pilgrimage literature of the Middle Ages, such as issues of genre, authorship, and originality, also apply to these two, which are, of course, dependent upon earlier descriptions,[4] but those problems will not be discussed in this study.

The cleric Ludolf von Sudheim traveled in 1336 to the Holy Land, where he then spent five years.[5] Written around the middle of the fourteenth century and later enjoying great popularity,[6] his travel description does not specify the length of his sojourn on Cyprus. However, he devotes care to an extensive and carefully structured description of the island, thereby allowing the readers to orient themselves easily. He begins the depiction with a short history of Cyprus and then adds a description of the most important cities, moving from west to east, providing a clear division into the three dioceses and one archdiocese: Paphus, Nymoni (or Limassol), Famagusta, and Nicosia, the latter described as closely connected to Constantia or Salamis. For the most part, city descriptions adhere to

---

grata temperie perfruantur, haec prope contra naturam intolerandis ardoribus aestuat, quasi hominum complexio ad elementa transierit. Noli ibi multum immorari. Non est enim militaris certe neque virilis habitatio. Fastus Gallicus, Syra mollities, Graecae blanditiae ac fraudes unam in insulam convenere. Quod optimum atque pretiosissimum habent: illic dissimillimis moribus aliunde veniens iacet Hilarion.

[3] Cf. K. S. Frank, p. 7f.

[4] Ganz-Blättler, p. 99f.

[5] Cf. also Stapelmohr, p. 5.

[6] Stapelmohr, p. 6. Concerning reception (as well as authorship), see also Ganz-Blättler, p. 47.

the following template: the history of the city in question and its relationship to the story of salvation, the present-day holy places, products of the city or region, and the riches available. Last, Ludolf provides general information about Cyprus, current personal reports, and information about the cities that had to pay tribute to the King of Cyprus.[7]

The Augustinian Jacopo da Verona describes the island differently. He was probably traveling on a diplomatic mission for a higher authority in 1335 or 1336. One point of temporal orientation is the siege and conquest of the city of Lajazzo (today the Turkish town of Ayas).[8] In his depiction, Jacopo provides a more personal picture with individual experiences. However, unlike Ludolf's, his portrayal was apparently seldom read and has been handed down in only a few manuscripts.[9] According to his own account, he spent twenty days on the island. In this description he first reports on his visit to King Hugo IV of Cyprus. His report contains many personal experiences and impressions of Famagusta, as well as general characteristics and points of interest on Cyprus. The two descriptions are – where necessary – supplemented by the description of Wilhelm von Boldensele, who had spent time on Cyprus in 1333,[10] and by the widely distributed description by Jean de Mandeville,[11] who, like Petrarch, was an armchair traveler in this case.

II. *Inertia* and *deliciae* on Cyprus: Indolence, Luxury and Venus

Petrarch's first reproach against the Cypriots is their indolence, their luxury, and their inclination to pleasures and emphasis on Venus.

There is nothing, however, about indolence or *inertia* to be found in the other travel descriptions, which concentrate on the climate that the people of Cyprus have to deal with. Jacopo da Verona, who was also there in late June, reports of this heat with a certain sympathy: *that the people can hardly keep themselves*

---

[7] Compare Ludolf, ed. Deycks, pp. 29-35; Ludolf, ed. Stapelmohr, pp. 105-113.

[8] F. Khull, p. 45 dates the trip later, similarly Röhricht, Deutsche Pilgerreisen, p. 46. Ganz-Blätter, p. 48, however, dates it as 1335. Concerning the conquest of Ayas, see Luttrell, p. 128f.

[9] Ganz-Blättler, p. 48 mentions only three manuscripts, a Latin one and two German ones from the fifteenth and sixteenth centuries.

[10] See Khull, p. 5f. See Wilhelm (Guillaume), ed. Deluz, p. 209f.

[11] Concerning the wide dissemination of the manuscripts, see Ganz-Blättler, p. 49. From the fourteenth through the sixteenth centuries, she indicates 50 French, 95 German (transl. by: Otto von Diemeringen, Michel Velser), 34 Latin, 31 English, 10 Italian, 5 Czech, 3 Danish, 2 Dutch, 2 Irish, and one Spanish manuscript. This does not even include the print versions; regarding these, see Bremer/Ridder, p. V-VIII.

*alive in summer and no one comes out of their homes except at night or in the
morning up until the third hour and from Vespers until nightfall.*[12] As shown by
this report, the inhabitants have adapted to the climatic conditions. This is
apparently also the case with dining customs, as Jean de Mandeville alone
mentions. Because of the extreme heat, both masters and servants use the
coolness of the ground. They dig an approximately knee-deep hole in a room,
where they then sit to eat – though never in the presence of guests.[13]

Luxury or *deliciae* primarily refers to the luxurious lifestyles or pleasures that
the rich upper class, nobility for the most part, could afford. Ludolf von Sudheim
reports extensively on their pleasures, summarized as *hunts, tournaments, and
other knightly games,*[14] but devotes particular attention to describing their
customs of the hunt, which differ from those with which he was familiar. These
customs are rooted in the extreme wealth and the lifestyle of many of the
nobility. Many of them had fled from the city of Accon after its fall in 1291 and
had found great riches or the opportunity to increase their own fortunes in
Cyprus – often at the expense of the indigenous Greeks, though Ludolf fails to
mention this point.[15] Ludolf reports that the nobles participate in a hunt for
inordinately extended periods – often as long as an entire month, carrying their
food on camels and sleeping under the stars. The hunts often involve formidable
expense. Many nobles can afford ten or eleven falconers or keep five hundred

---

[12] *Item, in insula illa Cypri, tantus calor est, ut vix homo in estate possit vivere, et nullus
exit domos nisi de nocte (et de mane usque ad terciam et ab hora vesperarum et
citra; fere ego fui mortuus in illo calore.),* Jacopo da Verona, ed. Röhricht, p.
178f.

[13] Mandeville, ed. Bremer/Ridder, Velser [20], Diemeringen [211]. Mandeville (Velser),
ed. Morrall, p. 19 (7rb): *In Cypren hond sie die gewunhait daz edel und unedel
essent uff der erde, und haissent machen gru(o)ben umb und umb ains knúß tieff,
und da legend sie scho(e)n dischtu(e)cher in. und wenn sie essen wellent, so
setzent sie sich dar in und tu(o)gent scho(e)ne dischlachen umb die gru(o)b. Und
da essent sie, das ist ir gewonhait. Und das tu(o)nd si daz sie dester ku(e)ler
habend, wann das land ussermassen haiß ist. Und wenn daz ist das yeman
fremder zu(o) in kumpt, den beraittend sie stu(e)l und disch und benck recht als
hie. Aber in gefelt ir syt basß.*

[14] *vnde hebben dar tiidkortinghe myt spele vnde myt torneye vnde menliken myt
iachtspele, wente dar is vele wyldes,* Ludolf, ed. Stapelmohr, p. 110f., Ludolf, ed.
Deycks, p. 33: *[Q]uotidie hastiludiis, tornamentis et specialiter venantionibus
insistentes.*

[15] For the situation of the native Cypriots, see Maier, p. 108f. Only a small segment
enjoyed such fabulous wealth, ibid., p. 111, cf. also Tzermias, p. 10f. Seldom is
the oppression of the indigenous people mentioned, as it is, for example, by Jacob
Wormser 217v, ed. Feyerabendt, in the sixteenth century. This aspect is usually
ignored.

dogs and a handler for each pair of them.[16] Ludolf reports this only as fact without passing any negative judgment.

In addition to the usual chase with dogs, Ludolf also mentions a special type of hunt, and Jean de Mandeville describes it as follows:

In Cyprus, they hunt with another animal in the same way we do with dogs. This animal looks like a leopard, is courageous and good-natured and attacks the prey boldly. It is a bit larger than a wolf and more bloodthirsty than a dog.[17]

This clearly indicates a cheetah, which can easily be tamed. The nobles from Outremer probably brought this form of the hunt to Cyprus along with their cheetahs and wanted to continue participating in the sport. The hunt with these unusually fast animals, which also hunt in small groups, seems to have impressed travelers such as Wilhelm von Boldensele as well.[18] Therefore it is not surprising that in many western European miniatures oriental rulers are also depicted accompanied by cheetahs.

The luxury that Cyprus can afford as an important center of trade affects daily life as well. Spices are as common as bread, as Ludolf describes. In one apothecary's shop one could find as much as four wagonloads of precious Aloe-wood, also used as incense.[19] The wealth of the people is legendary. Ludolf tells of a burgher's daughter, whose headdress is valued as more precious and sumptuous than that of the queen of France. Normal dimensions apparently do not suffice for describing the wealth of Cyprus. He does not want to report on more precious stones and other riches at the risk of seeming unbelievable.

---

[16] Ludolf, ed. Deycks, p. 34; Ludolf, ed. Stapelmohr, p. 111.

[17] See Mandeville, ed. Bremer/Ridder, (Diemeringen) [211]: *Jn Cypern iaget man mit einem tier als man hier tu(o)t mit hunden vnd ist das tier gestalt als ein Lebart vnnd ist frech. Vnd fro(e)dig vnd vallet die tier die man iagt freuelichen an vnd ist ein wenig gro(e)sser dann ein wolff vnd vast frecher vnd getürftiger dan ein hund. Doch iaget man ouch do mit hunden.* Compare also Velser's version in Anton Sorg's print edition, ibid. [20], where the description is accompanied by an illustration. Regarding the description, see also Mandeville, ed. Morrall, p. 19, (Velser-Übers.): *In Cypern ist ain tier, das haissent sie papions, und ist ainem leopard gelich und ist gro(e)sser denn ain lo(e)w. Da rittent sie mit an das geja(e)gt, wann es vahet allu tier und ist fast fraidig und jagt als ain hund.* (7ra). Concerning this kind of hunt, also Ludolf, ed. Stapelmohr, p. 111, or Ludolf, ed. Deycks, p. 33.

[18] Confer Deluz, p. 210.

[19] This probably refers here to aloe wood, which comes from a type of aquilaria and was prized for smoking. (*aloes lignum, xiloaloes, aloxilon, agallochum* etc.), Cf. I. Müller, p. 453.

Luxury and the "good life" seem ubiquitous, as Ludolf notes in an objective, if astonished, manner.[20]

Petrarch puts forward a strong case with the veneration of Venus, in which he refers to the past and, by metonymy, to the soft life of the Cypriots. According to Hesiod's "Theogony", Aphrodite, "born of the sea-foam," emerged from the sea on Cyprus. The "Odyssey" also names Cyprus, in particular Páfos, as her homeland.[21] Even in the fourteenth century, Venus and Páfos, as the place of her veneration in antiquity, are still vividly present in the consciousness of travelers and inhabitants. Ludolf reports that in former times, nobles, both men and women, came great distances to her place of veneration at Páfos. In antiquity, Aphrodite's sanctuary there in Páfos was also known as the "navel of the world".[22] And even in modern times she is not yet forgotten and is, in some respects, honored, for the church at the edge of the former sanctuary is dedicated to the Panagia Aphroditissa, the holiness of Aphrodite, and young mothers make offerings in front of a stone in the wall of the sanctuary.[23]

This place seemed a dangerous one to Ludolf as he intertwines past and present in his depiction. He refers to the ancient temple prostitution, which had influenced the present-day mentality of the inhabitants of the city, as well as all of Cyprus. In his estimation, they are the most self-indulgent people, and yet he excuses them by acknowledging that no one could possibly resist the seductive atmosphere of the place:

> *For on this island, and especially at the location of the Temple, if one were to lie down and sleep on the ground, the earth itself would entice him toward love and fornication the entire night.*[24]

According to the account of the history of the Trojan War Ludolf presents, it is no wonder that Helena is kidnapped by Paris while on the way to the Temple of Aphrodite in Páfos.

---

[20] Compare Ludolf, ed. Deycks, p. 32f., 35.

[21] Odyssey, 8,362-366. Demodocus sings about Aphrodite's adultery with Ares. Through Poseidon's appeal from Hephaistos' net, she turns back to Páfos. Felix Fabri, in the fifteenth century, seems to have been the first to have connected the Saga of the Venusberg to Páfos, thereby correcting the Tannhäuser myth as well as his contemporaries' localization of the Venusberg at Rome. Fabri, Evagatorium III, p. 221f.

[22] Concerning Páfos as a former place of venerations of Aphrodite, cf. Ludolf, ed. Deyck, p. 30. Regarding its meaning as the center of the world, see Schneider, p. 140.

[23] Schneider, p. 141f.

[24] Ludolf, ed. Deyck, p. 30. *Nam si terra Cypri et specialiter loci, quo castrum Veneris stetit, capiti dormientis supponeretur, ipsum ad libidinem et ad coitum per totam noctem provocaret.*

Ludolf adds that even the books of the Bible connect Cyprus with the concept of love, at least in the Vulgate's version of the "Song of Solomon", where in verse 1:14 the beloved is metaphorically described as *a cluster of Cyprus grapes to me, in the vineyards of Engaddi.*[25] Scholars including the Venerable Bede, Hrabanus Maurus, and Honorius Augustodunensis also connected Cyprus with this verse, although they thought less of the grapes, but rather of an anointing (nard) oil made from a particular tree.[26] Haimo of Auxerre suggests a possible location for Engaddi – not in Judea, but in connection with a vineyard of the same name in Cyprus – when he comments that *Cyprus insula est ubi vites nobiles esse feruntur, maximos botros procreantes/ Cyprus is an island where there reportedly are noble vineyards producing the largest grapes.*[27] Wine is another luxury that the Cypriots use to excess, for according to Ludolf one finds *nowhere in the whole wide world bigger tipplers than in Cyprus,* and this is the case even though the strength of the wine is so extraordinary that it has to be watered down at a ratio of nine to one to be consumed; otherwise grave consequences could result, for *if anyone drank a whole barrel full of this wine, he wouldn't become drunk from it, but it would burn out and destroy everything on the inside of his body.*[28] Ludolf is familiar with the production of this famous wine, which is also mentioned by Johann von Würzburg[29] and Heinrich von Neustadt.[30] This wine production is at least partially in clerical hands. The vineyard Engaddi at Páfos once belonged to the Knights Templar, who used captured Saracens to cultivate it. Since Ludolf can count on interest in the description of the wine, he describes wine and winegrowing rather extensively. He names the various varieties of grapes, the best places for planting on the island, and a phenomenon concerning the wine's color, which is at first red, but which, after years of storage, fades to white – with no loss of strength, however.[31]

---

[25] This quote refers to the verses in the Vulgate: *Botrus Cypri dilectus meus mihi, in vineis En-gaddi.* The verse was also indirectly cited by Wilhelm von Boldensele. Boldensele, ed. Deluz, p. 209.

[26] Beda Venerabilis, Allegoria Expositio in cantica canticorum IV,5,21, PL 91, 1164; Hrabanus Maurus, Commentaria in Ecclesiasticum V,14, PL 109, 936; Walafridus Strabo, Cantica canticorum IV ad versus 13, PL 113, 1151; Honorius Augustodunensi, De imagine mundi tract. I, I ad versus 13, PL 172,378.

[27] Haimo of Auxerre, Commentarium in cantica canticorum I, PL 117, 300.

[28] Ludolf, ed. Deycks, p. 34: *et si homo dolium plenum de vino illo biberet, ipsum non inebriaret, sed interiora eius cremaret et annihilaret.[…] et non sunt in mundo meliores et maiores potatores, quam in Cypro.* Concerning Engaddi, cf. also ibid, p. 30.

[29] Johann von Würzburg, Wilhelm von Österreich, vv. 363f.

[30] Heinrich von Neustadt: Apollonius von Tyrland, vv. 2765-69.

[31] Ludolf, ed. Deycks, p. 31. Regarding the Templars on Cyprus, cf. Maier, p. 101; Reden,

III. The Melting Pot of the Mediterranean and its Prominent Personalities

Petrarch's reproach that the island had produced none of its own famous men (*nunquam vir aliquis clarus*) is only partially true for Cyprus. Although Zeno is not named in the aforementioned travel descriptions, Hrabanus Maurus mentions the founder of the Stoa as coming from Cyprus.[32] And in the story of Christianity, Cyprus is not insignificant either. Jean de Mandeville names St. Zemoman and a St. Barbara as coming from Famagusta.[33] Ludolf knows another *virago* who has proven her virtue: St. Katherine, born in Salamis.[34] If one considers the spreading of the Christian message, Cyprus ranks ahead of even Rome as a focus of the preaching activity of Paul and Barnabas.[35] In addition, in the Monastery of the Holy Cross, today's Stavrovoúni monastery, founded by the Empress Helena, one can find the cross of the right-hand thief (with an iron nail from Christ's cross) that even Jacopo da Verona touched with devotion.[36]

Jacopo does not reproduce the mixture of negative characteristics of the island's inhabitants and their generalization that Petrarch lists with his polemical connection and mixing of Greeks and Franks and the influence of oriental lifestyle. Jacopo finds King Hugo IV of Lusignan to be a *virtuous, gracious and pious ruler*.[37]

Many travelers took an interest in the relationships of these many components, which also indicate a certain internationalism attached to Cyprus as an important trading center in the Mediterranean. This is also underscored by its multilingualism and the learning of foreign languages in Cyprus, also promoted by language schools, as Ludolf reports.[38] He also takes note of customs different from those in Western Europe. In Famagusta, Jacopo da Verona not only had the opportunity to observe the misery of the refugees from Armenia Minor fleeing

---

p. 49.

[32] Hrabanus Maurus, Commentaria in Jeremiam I, 2, PL 111, 813.

[33] Mandeville, ed. Ridder/Bremer, (Diemeringen) [211], (Velser) [20], Ludolf, ed. Deycks, p. 33.

[34] Ludolf, ed. Deycks, p. 33.

[35] Ludolf, ed. Deycks, p. 32.

[36] Jacopo da Verona, ed. Khull, p. 54, compare also Ludolf, ed. Deyck, p. 32. Concerning the Monastery of Stavrovoúni, regarded as the oldest on Cyprus, see Schneider, p. 78. The cross of the right-hand thief is also mentioned by Mandeville, ed. Bremer/Ridder, (Diemeringen) [211], ibid. (Velser) [19].

[37] Jacopo da Verona, ed. Khull, p. 51 or Jacopo, ed. Röhricht: *vertuosus ac gratiosus et Deo devotus*, p. 176. King Hugo IV died in 1359, von Reden, p. 195, Maier, p.117.

[38] Ludolf, ed. Deycks, p. 34, Ludolf, ed. Stapelmohr, p. 112: *Vortmer hort me in Cipro vnde spreket dar vn*de *leret in su*nderghen *scholen alle de sprake, dede werlt hat.*

from the Mamluks,[39] but also to witness indigenous rites performed in the burial of the dead or at weddings. The linguistic proximity of the orient is also evident in his description of the burial customs.

> *there sat two women at the head and two at the foot of the deceased, wailing with loud voices [...] and singing in Greek, so that I could not understand anything (for in all of Cyprus they speak Greek), however they understand Saracen and French, [...] for when I asked what they were singing, I was told that they were praising the deceased for his beauty, wisdom and strength, on account of his mercy and other virtues.*[40]

The connection to the former Latin Kingdom is alive in Cyprus and reminds one daily of the glorious, though long-lost, past.[41] Thus Jacopo describes an

---

[39] Jacopo, ed. Khull, p. 52. Jacopo shows a great deal of compassion in the description of the refugees and uses them intentionally to drum up support for a crusade, ibid. Jacopo reports on 1500 refugees who had fled from Lajazzo to Cyprus. *O Lord, what great sadness there was there, where one could see such great weeping and wailing, the old gray-haired men weak with hunger, the little children clutching the breasts of their mothers in the streets of Famagusta! The Christians, above all the princes, the nobles and the wealthy should hear this, they who lie idly in their palaces and cities, eating and drinking and satisfying all their cravings. There they pay no attention to winning the Holy Land nor to bringing it back to the Christian faith!/ O Domine Deus, quanta tristicia videre ipsam multitudinem cum planctu et ejulatu filios lactantes ubera in platea Famagoste ad pectora mulierum, senes, canes famelicos lamentantes; audiant hec cristiani, qui in suis civitatibus et domiciliis habitant, comedentes et bibentes et seipsos in deliciis nutrientes, qui terram sanctam non/ curant acquirere et ipsam ad cultum deducere cristianum,* Jacopo, ed. Röhricht, p. 177 (translation by Khull). King Peter I of Cyprus (1359-1369) tried to carry out the crusade that Jacopo had called for. His brilliant appeals to the courts of Europe were impressive, but in vain. See Maier, p. 118-121, von Reden, p. 194-203. The Cypriot songs mentioned by Sibylle von Reden, praising Peter I as "King of the East and Monarch of the West" and as a "romantic Lover" were not mentioned by later travelers, cf. ibid. p. 203. About the Armenian people of Lajazzo (Ayas) fleeing from the Mamluks, see Jacopo, ed. Röhricht, p. 177 and Luttrell, p. 128. The town was conquered in 1336, ibid. Regarding the refugees on Cyprus, see Maier, p. 117.

[40] Jacopo, ed. Khull, p. 52. Regarding the Armeniens, compare Röhricht in: Jacopo da Verona, ed. Röhricht, p. 178f.: *et ecce ad caput due mulieres cantantes alta voce, et due ad pedes pie lamentantes [...] et cantabant in lingua greca, et ideo eas non poteramus intelligere, quia omnes de Cypro loquuntur grecum, bene tamen sciunt saracenicum et linguam francigenam, [...] quesivi, quod dicerent; dixerunt, quod laudabant mortuum de pulchritudine, de providencia et de aliis virtutibus.* Loud lamentation of the dead is still common today; see von Reden, p. 268.

[41] This connection to the Kingdom of Jerusalem had existed since Hugo III (1267-84). The crowns of Cyprus and Jerusalem were united under his rule, as well as under

impressive wedding procession, but one whose joyful mood is subdued by recent events. The bride on horseback has painted eyebrows and forehead. Twenty-four burning wax candles are carried in front of and behind her. After the candle-bearers following the bride are women almost completely covered in black cloaks (reminiscent of the Islamic Burka), a fashion that Jacopo explains in more detail:

> *All respectable Cypriot women go around in such clothing outside their homes, so that one sees nothing of them but their eyes, and so all of them wear black, since the time the Christians lost the city of Accon or Ptolemais.*[42]

The conglomeration of the many Christian sects in Famagusta alone is impressive. Jacopo writes with short explanations about the Latins (that is, "true Christians"), Greeks (Greek Orthodox), Jacobians, Maronites, Armenians, Georgites, and Nestorians who can be found there.[43]

This plethora of denominations could also lead to original ecumenical attitudes. What is also noteworthy is a remark from the fifteenth century by Felix Fabri: he mentions an "ecumenical" priest near the Convent of the Holy Cross who holds masses in Latin as well as in Greek. He is Latin in clothing and appearance and calls himself a Barefoot Friar (Franciscan),[44] but like the Greek priests, he had a wife and child.

Henry II, King of Cyprus (1285-1324). After Accon and Tyre fell in 1291, Henry II continued to hold the title of King of Jerusalem. Famagusta became the nominal seat of the Kingdom of Jerusalem and here he had himself crowned King of Jerusalem. He appointed high-level dignitaries for the Kingdom of Jerusalem. J. Richard, p. 741. For a short summary of the history of Cyprus and its connection to Jerusalem, see Ludolf, ed. Deycks, p. 30.

[42] Jacopo, ed. Khull, p. 53 and Jacopo, ed. Röhricht, p. 178: *quia tali modo vadunt omnes domine de Cypro, et non videntur nisi oculi, et dum domos exeunt, semper cum clamide nigra vadunt, et a tempore, quo cristiani perdiderunt Acri, quod est Acon sive Ptolomayda.*

[43] Jacopo, ed. Khull p. 53; Röhricht, Rev. Lat, p. 178, see also Maier, p. 113 regarding the Christian sects on Cyprus.

[44] Felix Fabri, Evagatorium I, p. 177: *Didici autem, quod clericus ille erat monachus, quod tamen habitu cognoscere non potui, quia toga de schamlotta opertus erat, et fuit ambarum ecclesiarum plebanus, graecae et latinae, et per omnia se regebat secundum utrumque ritum.* However, Felix Fabri disapproved of this, as can be seen clearly from the statements that follow.

IV. Epilogue

A comparison of Petrarch's description of Cyprus with that of some of his contemporaries results in a rather different picture. Even though positive components are lacking in Petrarch's depiction because of his rather selective *virtus* concept, the function of the "Itinerarium" as an inner self-portrait and his concentration on the *patria*, the descriptions in the reports of others who wanted to make certain information available, were more open and positive. Even Jean de Mandeville tries to heighten the appeal of the island and a journey to it by referring to the lost city of Atlantis, which one passes on the sea voyage to Cyprus.[45] So despite personal and functional remarks such as the exhortation to a new crusade in light of the Armenian refugees, the descriptions of the contemporaries prove to be more realistic and open to differences and the exotic nature of the island and its life, which they describe in more detail. Therefore in addition to the exact location of Cyprus, its connections to antiquity and to the Christian message, and its geophysical problems such as earthquakes, they delve into the products of the country, the characteristic flora and fauna, and the inhabitants. Remarkably, no one mentions the origin of the name of the island, which is well known in sacral literature and, according to Flavius Josephus, is derived from the city of Kition on Cyprus, today's Larnaca. This city's name itself stems from the ancestor Chetimus (Kittim).[46]

The contemporaries who had seen Cyprus with their own eyes found the wealth of this island pulsing with life and its cosmopolitan bustle particularly fascinating. Ludolf also emphasizes its importance as a meeting place of Occident and Orient.

So there are also the richest merchants in Cyprus, which should be no surprise to anyone, since Cyprus is a country located at the farthest edge of Christendom. For all ships, no matter what they are carrying and where they come from, must first land in Cyprus and nowhere else, so all strangers and pilgrims come first to Cyprus. And daily, from morning till night, one hears new voices and news. In Cyprus you can hear and know what's happening in the entire world.[47]

---

[45] Regarding Atlantis, see Mandeville, ed. Bremer/Ridder, (Velser) [19], (Diemeringen) [210]. Velser's translation speaks of the city *Saterea* that sank with the island. Cf. also Mandeville, ed. Morrall, p. 18.

[46] Flavius Josephus, Antiquitates I,6.1.

[47] Ludolf, ed. Deycks, p. 34: *Item in Cypro sunt ditissimi mercatores et cives, et non est mirandum, quia Cyprus est terra Christianorum ultima, itaque omnia navigia parva et magna et omnia mercimonia, etiam quaecumque sunt et de quacumque parte maris veniunt, semper primum in Cyprum necessario veniunt, quod aliquatenus non possunt praeterire. Etiam omnes peregrinos de quibuscumque mundi partibus ad partes ultramarinas tendentes oportet venire in Cyprum et*

In the fourteenth century, therefore, the Cyprus myth is one of wealth and luxury, the characteristic features that travelers and pilgrims recognized when sojourning there. This myth, as one befitting Cyprus's economic zenith, remained vividly alive well into the fifteenth century, as shown by pilgrims such as Bernhard von Breydenbach and Felix Fabri.[48] Evidently, the only occasional mention of the Lusignans means they did not occupy the center of the travelers' attention. Although the Melusina myths took shape in the fourteenth century and the story of the founding of Lusignan by a snake-woman was written down in the middle of that century, the travelers of the fifteenth century give no indication of a connection of this dynasty to the myth, which held no place of importance in their perception of Cyprus.[49] However, Cyprus and its kingdom, on the other hand, lent substance to the Melusina myth. Thus Jean D'Arras in his Melusina novel (1393) uses the myth in order to authenticate the possessions of the Lusignans in Cyprus and Armenia.[50]

Located on the doorstep of the Orient and at the height of its economic prosperity, Cyprus, with its riches, was indeed impressive in the fourteenth-century world. It was also seen by the travelers as the turnstile and pivotal point between the worlds of the Occident and Orient, as a place of world commerce,[51] and of the super-rich of the times, who seemed to be living in a luxurious paradise. Cyprus's reputation as an island of indescribable luxury and wealth spread throughout all of Europe and continued for the next hundred years, but in the course of political events, it lost its luster during the sixteenth century.[52]

---

quotidie a solis ortu usque ad eius occasum ibidem audiuntur rumores et nova.

[48] See also Maier, p. 111; Bernhard von Breydenbach, Cr; Felix Fabri, Evagatorium III, p. 218f.

[49] In connection with Armenia, Jean de Mandeville as well as Hans Schiltperger mention the Saga of the Sperberburg (anchored in the Melusina myth), in which a fairy or lady will grant a favor if someone could spend three or seven nights awake in the castle, cf. Pastre, p. 123. Concerning the Sperberburg, see also Thüring von Ringoltingen, Melusine, chapter LVIIIf. Regarding the origin of the myth and its connection with the dynasty of the Lusignans, see Lecouteux, p. 80. The Lusignans of Cyprus's coat of arms had a red lion on a golden field and the Jerusalem cross. Moreover, from time to time a red lion on a silver field was combined, the coat of arms of the Kingdom of Armenia, cf. von Reden, p. 200.

[50] Pastré, p. 124, see also Harf-Lancner, p. 504; Backes, p. 13; Thüring von Ringoltingen, Melusine, chapter XXf.; Maier, p. 102f.

[51] Concerning the importance of trade as a source of wealth for Cyprus, which profited from intermediary trade and from the fall of the Crusader states, see Maier, p. 112f.

[52] The portrayal of the holy sites was preserved in the descriptions of the sixteenth century, however: The wine of the island made a special impression. The extraction of salt from the sea is also mentioned here, as well as by Felix Fabri. Fabri, Evagatorium I, p. 178.

Bibliography

I. Sources

Bernhard von Breydenbach: Peregrinationes. In: Bernhard von Breydenbach. Peregrinationes. Un viaggiatore del Quattrocento a Gerusalemme e in Egitto. Ristampa anastatica dell'incunabolo. Traduzione italiana e note di Gabriella Bartolini e Giulio Caporali. Prefazione di Massimo Miglio. Saggio introduttivo di Gabriella Bartolini. Rom: Vecchiarelli 1999, pp. [1]-[198] (a-pv[qij]).

Guillaume de Boldensele (1336): Liber de quibusdam ultramarinis partibus et precipue de Terra Sancta. Suivi de la traduction de Frère Jean le Long (1351). Ed. par Christiane Deluz. Diss. (masch.) Paris (Sorbonne) 1974.

Flavius Josephus: Jüdische Altertümer. Übersetzt und mit Einleitung und Anmerkungen versehen von Dr. Heinrich Clementz. Wiesbaden: Fourier 1993.

Francesco Petrarca: Reisebuch zum Heiligen Grab. Itinerarium ad sepulcrum Domini nostri Jesu Christi. Lat./Dt. Übers., hrsg. von Jens Reufsteck. Stuttgart: Reclam 1999 (=UB; 8888).

Fratris Felicis Fabri Evagatorium in Terrae Sanctae, Arabiae et Egypti peregrinationem. Ed. Cunradus Dietericus Hassler. Vol. 1-3. Stuttgardiae: Soc. Litt. Stuttg. 1843-49 (Bibliothek d. Lit. Ver. in Stuttgart; 2-4).

Heinrich von Neustadt: Apollonius von Tyrland. In: Heinrichs von Neustadt 'Apollonius von Tyrland' nach der Gothaer Handschrift, 'Gottes Zukunft' und 'Visio Philiberti' nach der Heidelberger Handschrift. Ed. S. Singer. Dublin/Zürich: Weidmann 1967, pp. 3-328.

Homer: Odyssee. Griechisch/Deutsch. Mit Urtext, Anhang und Registern. Übertragen von Anton Weiher. Einführung von A. Heubeck. Düsseldorf et al.: Artemis & Winkler 2007.

Jacob von Bern (1346-1347). In: Deutsche Pilgerreisen nach dem Heiligen Lande. Hrsg. und erläutert von Reinhold Röhricht und Heinrich Meisner. Berlin: Weidmann 1880, pp. 43-64.

Jean de Mandeville. Reisen. Reprint der Erstdrucke der deutschen Übersetzungen des Michel Velser (Augsburg, bei Anton Sorg, 1480) und des Otto von Diemeringen (Basel, bei Bernhard Richel, 1480/81). Hrsg. und mit einer Einleitung versehen von Ernst Bremer und Klaus Ridder. Hildesheim;

80

Zürich; New York: Olms 1991 (Deutsche Volksbücher in Faksimiledrucken, A; 21). [Velser, 1-182, Diemeringen 183-388]

Johanns von Würzburg *Wilhelm von Österreich* aus der Gothaer Handschrift. Hrsg. von Ernst Regel. Nachdruck. Dublin; Zürich: Weidmann 1970.

Le pèlerinage du moine augustin Jacques de Vérone (1335). Ed. Reinhold Röhricht. In: Revue de l'Orient latin 3 (1895), pp. 155-302 (reprint Brüssel 1964).

Ludolfs von Sudheim: Reise ins Heilige Land. Nach der Hamburger Handschrift hrsg. von Ivar von Stapelmohr. Kopenhagen/Lund: Levin & Munksgaard/ Gleerup 1937.

Ludolphi rectoris ecclesiae parochialis in Suchem. De Itinere Terrae sanctae liber. Nach alten Handschriften berichtigt. Hrsg. von Ferdinand Deycks, Stuttgart: Litter. Ver. 1851 (BLVS; 25).

Petrarch's Guide to the Holy Land. Itinerarium ad sepulchrum domini nostri Yehsu Christi. Itinerary to the Sepulcher of Our Lord Jesus Christ. Facsimile edition of Cremona, Biblioteca Statale, Deposito Libreria Civica, manuscript BB. 1.2.5. With an introductory essay, translation, and notes by Theodore J. Cachey, Jr. Notre Dame, Indiana: Univ. of Notre Dame Press 2002.

PL = Patrologia Latina. Cursus completus. Ed. Jacques-Paul Migne. Paris: Garnier 1844-1865.

Sir John Mandevilles Reisebeschreibung. In deutscher Übersetzung von Michel Velser. Nach der Stuttgarter Papierhandschrift Cod. HB V 86. Hrsg. von Eric John Morrall. Berlin: Akademie 1974 (DTM; 64).

Zweier deutscher Ordensleute Pilgerfahrten nach Jerusalem in den Jahren 1333 und 1346. Nach ihren eigenen Aufzeichnungen erzählt von Felix Khull. Nebst einer Beigabe: Beschreibung des Heiligen Landes durch Johann von Würzburg. Graz: Styria 1895. [Wilhelm von Boldensele, pp. 7-46, Jacopo da Verons, pp. 47-105]

II. Secondary Literature

Backes, Martina: Fremde Historien. Untersuchungen zur Überlieferungs- und Rezeptionsgeschichte französischer Erzählstoffe im deutschen Spätmittelalter. Tübingen: Niemeyer 2004.

Boase, Thomas Sherrer Ross (ed.): The Cilician Kingdom of Armenia. Edinburgh; London: Scottish Academic Press 1978.

Buschinger, Danielle/Spiewok, Wolfgang (eds.): Melusine. Acte du Colloque du Centre d'études Médiévales de l'Université de Picardie Jules Verne, 13 et 14 Janvier 1996. Greifswald: Reineke 1996.

Feldbauer, Peter/Morrisey John: Weltmacht mit Ruder und Segel. Geschichte der Republik Venedig 800-1600. Wien: Magnus 2004.

Frank, K.S.: Hilarion von Gaza. In: Lexikon des Mittelalters. Ed. Norbert Angermann/Robert-Henri Bautier et al. Vol. 5. München: Lexma 1991, p. 7f.

Ganz-Blättler, Ursula: Andacht und Abenteuer. Berichte europäischer Jerusalem- und Santiago-Pilger (1320-1520). Tübingen: Narr 1990.

Harf-Lancner, Laurence: Melusine: In: Lexikon des Mittelalters. Ed. Norbert Angermann et al. Vol. 6. München: Lexma 1993, pp. 504-506.

Harf-Lancner, Laurence: L'image et le monstrueux: Geoffroy la grand dent, le sanglier de Lusignan. In: Buschinger, Danielle/Spiewok, Wolfgang (eds.): Melusine. Acte du Colloque du Centre d'études Médiévales de l'Université de Picardie Jules Verne, 13 et 14 Janvier 1996. Greifswald: Reineke 1996, pp. 77-92.

Kretschmayr, Heinrich: Geschichte der Republik Venedig. 3 Vol. Reprint Aalen: Scientia 1964.

Lecouteux, Claude: Zur Entstehung der Melusinensage. In: ZfdPh 98 (1979), pp. 73-84.

Löher, Franz von: Cypern in der Geschichte. Berlin: Carl Habel 1878.

Luttrell, A.T.: The Hospitallers' Interventions in Cilician Armenia: 1291-1375. In: The Cilician Kingdom of Armenia. Ed. T.S.R. Boase. Edinburgh; London: Scottish Academic Press 1978, pp. 118-144.

Maier, Franz Georg: Cypern. Insel am Kreuzweg der Geschichte. München: Beck 1982.

Müller, I: Aloe. In: Lexikon des Mittelalters. Ed. Robert Auty und Norbert Angermann. Vol. 1. München: Lexma 1980, p. 453.

Nicolau-Konnari, Angel/Schabel, Chris (eds): Cyprus. Society and Culture 1191-1374. Leiden: E. J. Brill 2005.

Pastré, Jean-Marc: Mélusine et les récits de voyage: le château de l'épervier chez Hans Schiltperger. In: Buschinger, Danielle/Spiewok, Wolfgang (eds.):

Melusine. Acte du Colloque du Centre d'études Médiévales de l'Université de Picardie Jules Verne, 13 et 14 Janvier 1996. Greifswald: Reineke 1996, pp. 123-134.

Reden, Sibylle von: Zypern. Vergangenheit und Gegenwart. 8000 Jahre Geschichte im Schnittpunkt dreier Kontinente. 2. überarb. und erg. Aufl. Köln: DuMont Schauberg 1974.

Röhricht, Reinhold/Meisner, Heinrich (eds.): Deutsche Pilgerreisen nach dem Heiligen Lande, Berlin: Weidmann 1880.

Richard, J.: Zypern. B. Das Königreich der Lusignan. Im späten 13. und 14. Jahrhundert. In: Lexikon des Mittelalters. Ed. Norbert Angermann et al. Vol. 9. München: Lexma 1998, p. 741f.

Schneider, Andreas: Zypern: Archäologische Schätze, byzantinische Kirchen und gotische Kathedralen im Schnittpunkt der Kulturen. DuMont. Kunst-Reiseführer. Köln: DuMont 1997.

Thierry, Jean-Michel: Armenien im Mittelalter. Translated from the French by Hermann Goltz. Regensburg: Schnell und Steiner/Saint-Léger-Vauban: Zodiaque 2002.

Tzermias, Paulos N.: Geschichte der Republik Zypern. Mit Berücksichtigung der historischen Entwicklung der Insel während der Jahrtausende. 4. überarbeitete und aktualisierte Auflage Tübingen et al.: Francke 2004.

Translation: Kandace Einbeck

Dr. Maria E. Dorninger
Institut für Germanistik
Akademiestr. 20, A - 5020 Salzburg
maria.dorninger@sbg.ac.at

Insulae: Myths, *Mujeres*, and Mexico

The Yucatán Peninsula, the thumb-shaped limestone shelf that protrudes out of southern Mexico into the Gulf, pointing roughly at Cuba and Florida, has, about five miles off the coast at its northeastern point, several elongated islands, one of which is the Isla Mujeres (The Island of Women). The island today sports a tourist village in the north, a landing strip in the middle, and a Maya temple cum lighthouse on its southern tip. Farther down the Yucatán coast, to the south, lies the much larger and culturally more significant island of Cozumel. The Isla Mujeres, however, differs from Cozumel in that its name, coming from the language of the Spanish colonizers rather than from the native Mayan language, derives from one of the very first encounters of a European colonial narrative with a native mythological complex. Within this encounter, the European expectations of the character of islands in the western Ocean collide with a set of Maya lunar and gynecological associations. The Isla Mujeres thus hosted, in post-colonialist terms, a clash of imaginaries.

Early Explorations

The extant sixteenth-century sources on the conquest contradict each other to a greater or lesser degree, but sifting out the probable from the chaff, one can reconstruct the following events. Before 1517, at least three expeditions had entered the general area: one, in 1502, headed by Columbus himself, had ventured into the Gulf of Honduras and ascertained the presence of land to the south of Cuba; the second, in 1508, by Yañez Pinzon, Díaz de Solis, and Ledesma had sailed along the eastern coast of Yucatán without making landfall; and in 1513 Ponce de León reportedly landed briefly on the north coast of Yucatán on his way back from his first Fountain of Youth search. More importantly for our topic, in early 1517, an expedition under Francisco Hernández de Córdoba left Cuba with three ships and went ashore at Cape Catoche, the tip of the thumb, as it were. According to Bernal Díaz del Castillo's eyewitness account, the Spaniards found there "a small square with three houses built of masonry, which served as *cues* or prayer-houses. These contained many idols of baked clay, some with demons' faces, some with women's, and others equally ugly which seemed to represent Indians committing sodomy with one another" (19). The temple also contained objects of poor-quality gold, which seems to have aroused the interest of Díaz's shipmates to a greater degree than had the idols. At this point no mention has been made of an island, nor of the naming of an island, nor has the author placed any real emphasis on the female figures, ranking them second after the demons.

The following year, 1518, Díaz accompanied Juan de Grijalva on an expedition which, this time, approached Yucatán from farther east and made first contact with Cozumel. Grijalva and his crew appear to have given Cape Catoche and environs a complete pass on their way to Campeche and thus have no bearing on Isla Mujeres.

Two years after the Hernández expedition, in 1519, the indefatigable Díaz found himself on the ship of Hernán Cortés, an expedition which would eventually bring all of Mexico under Spanish rule. After landing first at Cozumel and spending a week there, Cortés next made landfall at Cape Catoche, where, finally, Díaz explains the *mujeres* name: at Cape Catoche they found "four of their [the Maya's] *cues* or temples, and many idols in them, most of them very tall female figures. We called the place therefore the Punta de las Mujeres [the Cape of Women]" (66).

Yet again, Díaz gives us no indication that the temples were located on an island; in fact, internal evidence speaks against an island location. While on Cozumel, Cortés had rescued Jerónimo de Aguilar, a Spanish cleric who had, as the result of a shipwreck, spent eight years among the Maya on the Yucatán mainland, living as a slave. On one occasion he had had to carry firewood for twelve miles to another village and had collapsed under the load. The four temples on Cape Catoche, reports Díaz, lay in the village to which Aguilar had been ordered to bring the firewood. This anecdote anchors the events on the mainland and excludes the possibility of the Punta de las Mujeres lying on an island.

Despite the nearly half a century between the events and Díaz's recording of them, his credibility in this matter is enhanced by the other details he provides; in addition to transmitting – as closely as Spanish orthography would permit – the Mayan word "k'u" for 'holy place,' he also gives a reasonably accurate description of a Maya temple E-group. Díaz later spent most of his time in what is now Guatemala, as governor of Antigua; one might argue that his memory of the first contact with the Yucatán appears to have been reinforced rather than diminished by his subsequent immersion in Maya culture, as he undoubtedly encountered many such temples in his lifetime.

Cartography

This pair of encounters with Maya shrines and female clay idols firmly associates the Mujeres name with the area around Cape Catoche. The earliest extant map of the area, that of Peter Apianus in his 1520 *Cosmographia* places the label *punta de magieles*[1] (with *punta* abbreviated as *p:*) near the cape and consequently also near the adjacent island. Subsequent maps (e.g. an anonymous one from 1524) bear the legend *isla de mueres* (with *isla* abbreviated as *y.*) (Saville 447). If, as seems likely from Díaz's report, the *punta* was not part of the *isla* to begin with, the naming of the island appears to have resulted from a misinterpretation of a *p:* (for *punta*) as a *y.* (for *ysla*), and the designation *de mujeres* wandered from the mainland cape over to the nearby island. Perhaps the temple at the southern tip of the island did indeed once house female figures – as will become likely in what follows – but the application of the name to the island appears to have taken place on the desk of a European cartographer rather than having a direct origin in the Spanish encounter with Maya culture.

The circumstances of the island's naming are, admittedly, a minor historical point. Of greater interest are the cultural expectations which underlie and motivate the renaming.

Cultural and Archaeological Evidence

If, in good post-colonialist manner, one begins with the indigenous narrative, that is, the place-value of coastal temples and islands in Maya culture, one must immediately distinguish between, on the one hand, the southern Maya culture which dominates our knowledge of the Classical period (traditionally 250-800 AD, but the early date has receded considerably in the past decade) and, on the other, northern Maya culture, whose geographical extent roughly corresponds with the Yucatán and which blossomed in the post-Classic era. The Classic Maya rarely encountered islands, at best those in lakes and rivers, and thus Classic Maya iconography and mythology show little interest in islands. Much the same applies, in general, to the Yucatecan Maya, whose home is an immense limestone shelf which, thanks to its porosity, allows no rivers – and hence no inland islands – to form. The northern Maya depended – and depend today – on sinkholes in the limestone ("cenotes") to collect water. Naturally, the major cities grew around these sinkholes, such as the famous one at Chichén Itzá, meaning that settlement tended toward the interior of the peninsula, and that the great bulk of internal trade took place over a network of intercity highways known as *sacbeo'ob*, rather than on waterways. Long-distance sea trade, however, made

---

[1] *magieles* 'magicians' seems to have resulted from a misreading of *mujeres*.

use of a series of coastal towns or island settlements, notably on Cozumel. Such settlements, a few of which, like Tulum, turned into large cities, tend to manifest late Maya characteristics such as extensive fortifications and signal towers (Sabloff 27). These walled cities mark a cultural boundary with the Other, in this case probably Tabascan or Caribbean traders who plied the coasts with merchant ships. The differing degree of hospitality with which the Spaniards were received during their first expeditions has been tied to the importance of their landing sites as trade centers and thus to the inhabitants' openness to strangers: the trading island Cozumel was friendly; on the other hand, Campeche, where Hernández received his fatal wounds, was not (Sabloff 11, 28).

But the prevalence of small temples along the coastal islands and, especially, along the eastern coast of Cozumel, leads us into another complex having to do with the character of the temples. Here we turn to Fray Diego de Landa, who devoted much energy to understanding Maya culture, only to eradicate much of it in a book-burning in 1562. Landa's version of the Isla Mujeres naming event, in his 1566 *Relación de las Cosas del Yucatán*, tells us that Hernández "reached the Isla de Mujeres and that it was he who gave it this name because of the idols which he discovered there to the goddesses of that land such as Aixchel, Ixchebeliax, Ixbunic, and Ixbunieta; these were dressed from the waist down and had their breasts covered [*cubiertos los pechos*] in the Indian manner" (Landa 34). Here, archaeological evidence comes to our aid; the east coast of the peninsula does indeed demonstrate a cult of Ix Chel (Lady Rainbow) (Milbrath 147). Mayanists are still in disagreement about some of her attributes and homologues in other Mexican cultures, but Ix Chel appears to represent the Old Moon Goddess, a symbol of the waning moon. Pointing to mural paintings at Tulum as evidence, Susan Milbrath attributes Ix Chel's cultic presence along the east coast of Yucatán to the waning moon's "especially dramatic disappearance in conjunction, dipping into the waters of the Caribbean" (147-148). The motion of dipping into the ocean ties her to Muzencab, the so-called "diving god," who has features of a honey bee and who appears everywhere one looks at Tulum. As Ix Chel also appears in the context of honey-making, a wide-spread industry in Yucatán, she and Muzencab represent two gendered sides of the same attribute. In addition to her lunar and apiary qualities, Ix Chel also acts as sponsor for weaving (a decidedly female occupation in Maya lands) and childbirth. According to Milbrath, a pilgrimage route for pregnant women ran through Tulum and across the water to the island of Cozumel, which lies immediately across from it (148). On Cozumel, archaeologists have found evidence for an Ix Chel cult ranging from small clay figurines to a full-sized ceramic figure through which a hidden priest could speak to supplicants in the goddesses' voice (Freidel in Sabloff 107-113).

In sum, then, for the Maya, the coastal islands served as the easternmost vantage point for watching the descent of the Old Moon Goddess over the water, a source of particularly moon-sanctified honey (of which Cortés took a sample for the King of Spain), and a sanctuary for giving birth. Similar childbirth sanctuaries, such as Los Sapos near Copán, Honduras, are known to have existed high in the mountains surrounding Classic Maya sites. In mountainless Yucatán, the nearest equivalent of the isolated mountain refuge lay just off the coast, on islands. Undoubtedly, not only *punta de las mujeres*, but most of the coast and coastal islands had an abundance of temples with female figures at the time of the conquest. The presence of towers on many of the temples, as well as their placement away from population centers, implies a secondary function as a lighthouse or signal tower, such as one sees at the temple on Isla Mujeres today.

This gives us an idea of what the Spaniards encountered on the Yucatán coast and adjacent islands. The next question concerns their mode of interpretation, what José Rabasa calls "reading the islands" (60). Faced with objects of an unknown semiotic system, the conquistadors and, later, their cartographers could only read them with tools from their own cultural baggage in order to make them align with European expectations.

Dangerous Women

Based on their previous 25 years' experience in the Caribbean and 2000 years of mythology, the Spaniards expected to encounter islands as they ventured south and west. It took, for example, a decade after Hernández's expedition for the realization to sink in that the Yucatán peninsula was in fact *not* the island as which it appears on maps well into the 1540s. In the context of the first half of the sixteenth century, then, the distinction between the peninsula and the Isla Mujeres is a small one, as both figured as islands in the European consciousness. But just as the expectation of finding mainland China nearby was still very much alive in 1519, so, too, was the expectation of encountering the islands known from Biblical and Classical texts.

Columbus, after all, had attempted to identify Biblical sites along the coast of Central and South America, even going so far as to speculate that he had stumbled across the island of Eden as he compared the geography of Venezuela with the description in his Bible (Morison 150-151). Proud of his erudition, Columbus repeatedly refers to his bewilderment as lands appeared before him that had no correlatives in Latin or Greek literature (quote in Rabasa 76), especially after he *had* discovered on his second voyage that the island of Guadeloupe housed Amazons (Magasich 108). Columbus's remarks should sensitize us to the major frame of reference for explorers and cartographers of the early sixteenth century, that of classical mythology and, to a far lesser extent, the Bible. The dominant cosmography allowed only for islands in the western ocean, and in the colonial imagination a significant number of these islands was populated by women, especially dangerous women. Even Mediterranean or Near Eastern islands of which the classical myths report, but which had not yet been found, could possibly turn up, complete with female occupants, in the unexplored spaces of the ocean. Thus the likelihood of encountering Sirens, Amazons, Circe and Calypso, Gorgons, and Hesperides could not absolutely be discounted as one sailed farther and farther west. In Oviedo's 1535 *Historia general y natural de las Indias*, for example, we find the argument that the islands discovered by Columbus were, indeed, the Hesperides, although Oviedo's object has more to do with politics than culture, since the twelfth king of Spain was named Hesperus, and if the islands bear his name, they must have belonged to Spain for centuries already (17). For Peter Martyr d'Anghiera and Bartolomé de las Casas, the Hesperides were coextensive with the Cape Verde islands off the coast of Africa; but the absence of nymphs and trees with golden apples and guardian dragons on the Cape Verde islands meant that the fabulous elements could exist elsewhere on as yet unexplored islands. As Rabasa points out, the very vagueness of the fables lends itself to an elastic interpretation of location; since the Sirens had not turned up in the by now thoroughly explored Mediterranean, perhaps they existed beyond the Pillars of Hercules (60).

One can follow the process of mapping legend onto geography as the explorers head west; Antilia figured in fifteenth-century Atlantic lore as a large, albeit mythical, rectangular island just beyond the known world, and its name came finally to rest on the modern Antilles. So too with the Amazons; no one had as yet found their mythical queendom and its island, variously placed in Scythia, the Black Sea, or Asia Minor. So when Orellana, while exploring the Marañón river in Brazil, encountered a tribe of large, armed women, he immediately dubbed them "Amazons." Of the three names for the river – the Orellana, the Marañón, and the Amazon – we see which one stuck in the cartographers' imagination.

In a broader sense, the expectation of meeting dangerous women on islands correlates with a general feminization of the American hemisphere. Rabasa has documented at length the allegorical representation of America (a feminine noun of the first declension) as a slumbering female nude awakened by the clothed male European (23-47). Without becoming excessively Freudian, one can view the first quarter century of the conquest of the Americas as a conquest of *insulae*, themselves feminine and first declension, and thus as a quasi-sexual exploit such as might excite the imagination of a conquistador. This train of thought leads us into the role of women as translators, mediators, and – in the context of conquest – panderers for the Spaniards; but that takes us far from the topic of islands.

In sum, the Isla Mujeres and the circumstances of its naming give us a peek into the mapping of European cultural expectations of dangerous women and dangerous islands onto an indigenous narrative of waning moons and childbirth on peripheral islands. This European reading of Maya myth, plus, probably, a cartographer's misreading of a caption led to the creation of an Island of Women on the seam between Old World and New.

Bibliography

I. Sources

Díaz de Castillo, Bernal: The Conquest of New Spain. Translated by J. M. Cohen. London: Penguin Books, 1963.

Landa, Diego de. The Maya: Diego de Lands's Account of the Affairs of Yucatan. A.R. Pagden, translator. Chicago: J. Phillip O'Hara Inc., 1975.

Oviedo y Valdés, Gonzalo Fernández de: Historia general y natural de las Indias. Volume 1. Madrid: Ediciones Atlas, 1959.

II. Secondary Literature

Magasich-Airola, Jorge and Jean-Marc de Beer: America Magica: When Renaissance Europe Thought it had Conquered Paradise. Translated by Monica Sandor. London: Anthem Press, 2006.

Millbrath, Susan: Star Gods of the Maya: Astronomy in Art, Folklore, and Calenders. Austin: University of Texas Press, 1999.

Miller, Mary and Karl Taube: An Illustrated Dictionary of the Gods and Symbols of Ancient Mexico and the Maya. London: Thames and Hudson Ltd., 1993.

Morison, Samuel Eliot. Christopher Columbus: Mariner. New York: Plume, 1983.

Rabasa, José. Inventing America: Spanish Historiography and the Formation of Eurocentrism. Norman, Oklahoma and London: University of Oklahoma Press, 1993.

Sabloff, Jeremy and William L. Rathje: A Study of Changing Pre-Columbian Commercial Systems. Cambridge, Massachusetts: Harvard University, 1975.

Saville, Marshall H.: "The Discovery of Yucatan in 1517 by Francisco Hernandez de Cordoba." Geographical Review, Volume 6, Number 5 (November 1918), 436-448.

Prof. Dr. James Ogier
Department of Foreign Languages
Roanoke College
Salem, VA 24153, USA
ogier@roanoke.edu

Far Eastern Islands and Their Myths: Japan

I

A myth of an earthly paradise is found in every area of the world and, naturally, also in Japan. The tale of Sukuna-hikona provides a good example of the myth. Sukuna-hikona is a Japanese god and, according to the records of the age of the gods in the *Kiki*, he constructed the countries of ancient Japan.[1] His most well-known work is the construction of the country of Izumo with Okuninushi-no-mikoto or Onamuchi-no-kami (another name of Okuninushi), the main god of the country of Izumo, today's Shimane Prefecture. Afterward he went--or returned--to "Tokoyo," literally 'the eternal land,' an earthly paradise. Since then he is believed to live there. His "Tokoyo" is an original Japanese creation, independent of Chinese Taoist concepts of immortality.[2]

The *Kiki* and the *Man'yo-shu* tell of Sukuna-hikona.[3] The *Kiki*, that is, the *Kojiki* and the *Nihon-shoki* or the *Nihongi*, are the historical documents of ancient Japan including myths and legends. They were compiled in the early half of the eighth century. The *Nihon-shoki* contains the official historical documents ordered by the imperial command, and while the *Kojiki* was compiled by Ono-Yasumaro privately, it was done so also at the command of the Emperor. The literal meaning of the *Man'yo-shu* is the "Collection of Ten Thousand Leaves," that is, the anthology of the oldest existing Japanese poems.

According to the records of the age of the gods in the *Kiki*, a section of which deals with the Crown Prince (the Emperor Ojin, a legendary figure in the fifth

---

[1] The records of the age of the gods: in "Kami-tsu-maki" (Vol. 1) of the Kojiki and in "Kamiyo no kami no maki" (Part 1) of the Nihon-shoki. Kurano (1958) 50-147; Sakamoto (1967) 76-133; Philippi (1968) 45-159; Aston, Nihongi (1972) 1-63; Chamberlain, Kojiki (1981) 15-159.

[2] Orikuchi, Tokoyo (1965) 3-15, esp. 11-2; Asakura (1997) 257-8, 332.

[3] Kurano (1958) 106-9; Sakamoto (1967) Bk. I, 128-9, 130-2; Philippi (1968) 115-7; Aston, Nihongi (1972) 128-9; Chamberlain, Kojiki (1981) 103-6; Takagi, Vol. 1 (1957) 168-9, no. 3-355; Takagi, Vol. 2 (1959) 154-5, no. 6-963; Takagi, Vol. 4 (1962) 286-9, no. 18-4106; Kojima, Man'yo-shu-1 (1971) 238, no. 3-355; Kojima, Man'yo-shu-2 (1972) 152, no. 6-963; Kojima, Man'yo-shu-4 (1975) 269-71, no. 18-4106.

92

century) and his mother Empress Jingu, Sukuna-hikona was mentioned as follows:

17th day. The Prince Imperial returned from Tsunoga [Tsuruga, today's Fukui Prefecture]. On this day the Grand Empress gave a banquet to the Prince Imperial in the Great Hall. The Grand Empress raising her cup wished long life to the Prince Imperial. Accordingly she made a song, saying:

> This august liquor
> Is not my august liquor:
> This prince of liquors
> He that dwells in the Eternal land
> Firm as a rock—
> The august God Sukuna,
> With words of plenteous blessing,
> Blessing all around—
> With words of divine blessing
> Blessing again and again—
> Hath sent as an offering to thee.
> Drink of it deeply.
> Sa! Sa![4]

The Empress Jingu, while waiting for the Crown Prince Ojin's return, composed the poem "Machi-zake" ("zake" is a variant of "sake," Japanese wine). "Machi-zake" means the specially brewed sake associated with waiting for a precious and valued person who is to return or visit.

As the song by the Empress reveals, Sukuna-hikona was believed to bring sake, medicines, and miraculous good things to the people of ancient Japan, and thus came to be regarded as the god of sake and fertility. In Japan, sake is assumed to be a miraculous remedy for all sorts of maladies, and thus Sukuna-hikona is also regarded as the god of medicine.

The literal meaning of "Tokoyo" is 'the eternal land.' The record suggests that "Tokoyo" is an ideal wonderful land, the earthly paradise.[5] At first, the conception of eternal life or eternal youth was not found in the idea of "Tokoyo." However, a conception of an earthly paradise influenced by Chinese thought was introduced from China.[6] Gradually, the foreign concept came to be mingled with the native one, and "Tokoyo" came to be regarded as the same land as Mt. P'eng-lai, which belongs to Taoist immortality concepts in China. For this

---

[4] Trans by Aston, Nihongi (1972) 244. Cf. Philippi (1968) 270-1; Chambers, Kojiki (1981) 297.

[5] Orikuchi (1965) 3-15, esp. 9-11; Tanigawa (1989) 1-286, esp. 198-203.

[6] Orikuchi (1965) 11-2.

reason, the Chinese ideograph of "Horai," pronounced as "Mt. P'eng-lai" in Chinese, reads "tokoyo" according to the traditional Japanese pronunciation in the *Nihon-shoki* and in other ancient Japanese documents.

By considering the idea of "Tokoyo," the ancient Japanese conception of the eternal land and eternal life can be brought to light. In this paper we will consider the myths as well as the literary narratives, while also taking Chinese historical documents into consideration.

II

According to the *Shih-chi* of Ssu-ma Ch'ien, the historical documents of ancient China, the Emperors believed in three holy mountain islands – Mt. P'eng-lai, Mt. Fang-chang, and Mt. Ying-chou – where the immortals lived.[7] They were said to possess the elixir of life.

The *Shih-chi* tells that the First Emperor of Ch'in, around the third century B. C., was famous for his eagerness to acquire the elixir of life and that Hsü Fu, Fang-shih, a benevolent wizard, went out to sea to search for medicine by order of the Emperor.[8]

Hsü Fu could not find the elixir of life and was afraid of the Emperor's great ire. So he told a lie and said that, although he had found the elixir of life, a big fish had been in his way and prevented him from reaching the island. Then the Emperor himself killed the fish, and Hsü Fu went out again. While waiting for Hsü Fu's return, however, the Emperor fell sick and died.

Another record in the *Shih-chi* tells that *Ch'in Shih-huang-ti* ordered his vassals to get the elixir of life.[9] At that time, the Emperor believed that the immortals might grant him the medicine if he offered a lot of gifts. He decided to offer boys and girls to the "Sennin" people. Then, those who had to sail out into the sea told a lie together and reported that Mt. P'eng-lai had slipped away or sunk into the sea or hidden in the mist whenever their boat was about to reach the shore. The Emperor again and again tried in vain to lay hold of the medicine and died.

Another record also tells of Hsü Fu, who told the Emperor that the main god of P'eng-lai had not allowed the elixir of life to be given to the Emperor because he had not dedicated enough offerings to the god.[10] It was a lie but the Emperor

---

[7] Cheng (1934) 142, 153-55; Noguchi (1968) 78, 83-5, 274-5; Honda (1969) 317, 331, 496.

[8] Cheng (1934) 142, 153-55; Noguchi (1968) 239; Ogawa (1975) 246-7; Ssu-ma (1982) 247-8, 258-9, 263-4, 1369-7: 3086; Tsuji (2004) 82-9.

[9] Noguchi (1968) 274-5.

[10] Noguchi (1968) 239.

ordered Hsü Fu to start for Mt. P'eng-lai again with three thousand men and women, and a good many workmen with various grains for the offerings. Hsü Fu went away with them. But on the way he discovered a spacious field and became king there.

The legend of Hsü Fu is narrated also by Chinese poets, for example, by Pai Chüi, a great poet in the T'ang period from the eighth to the ninth century. The poet, criticising the politics of the Emperor who had annoyed so many people, in his poems caused Hsü Fu to age in the boat.

> The sea endless and brimming with water
> The wind blowing all over
> Eyes looking for the island eagerly,
> The crews cannot find Mt. P'eng-lai,
> Not finding Mt. P'eng-lai,
> They could return by no means.
> Young boys and girls grew older in vain in the bark,
> Hsü Fu made compositions and his words are full of lies.[11]

The legends of Hsü Fu were found also in several districts of Japan.[12] According to the legends, Hsü Fu reached the seashore of a village as good as dead. He was saved and taken care of by the villagers. One story tells that, despite his once being saved, in the end he died.[13]

Hsü Fu was also mentioned in classical Japanese literature, which was in its first stage, influenced by Chinese literature, especially by Pai Chüi.[14] Accordingly, Japanese narratives of the period picked up this image of him growing old in a boat.[15] For example, in the "the Book of a Butterfly" of the *Tale of Genji*, the image of Pai Chüi is thus quoted.

Never again will I dream of the Mountain on the Tortoise's Back, for here in this boat have I found a magic that shall preserve both me and my name forever from the onset of mortality.[16]

The "waka" poem was composed when Hikaru Genji, the protagonist of the *Tale of Genji*, enjoyed a boating party with the other nobles. The poet emphasises that it is unnecessary for them to search for Mt. P'eng-lai on the

---

[11] Pai Chüi. Trans. by the writer. Text; Saku (1978) 246. Cf. Saku (1978) 245-47; Tsuji (2004) 89-102.

[12] Cheng (2000) 201-68; Tsuji (2004) 30-56,103-48.

[13] Tsuji (2004) 39.

[14] Tsuji (2004) 89-102.

[15] Tsuji (2004) 92-102.

[16] Trans. by Wailey (1927) 220. Cf. Abe (1972) 159; Taylor (2003) 442.

tortoise's back. The image is based on Pai Chüi's poem. The party members would not like to grow old in the boat but rather to enjoy the time of the party to their hearts' content.

Traditionally it has been assumed that the island of Mt. P'eng-lai is held up by a tortoise (or tortoises) in ancient Japan. Wayley translated the tortoise in the singular, but, according to other narratives, the number of tortoises is plural.[17] In this case, the tortoises do not carry the island on their backs but rather they hold it up with their heads.

III

The *Kiki* do not tell of Hsü Fu at all but relate an episode similar to the tale of Hsü Fu. The record may possibly imply that the Emperor Suinin was keenly longing for eternal youth and long life. The Emperor was the eleventh Emperor and said to be a legendary figure. According to the *Kiki*, he once ordered Tajima-mori, his vassal, to fetch the fruit of "Tokijiku-no-kaku-no-mi" in the land of "Tokoyo." He started at once and, after many hardships and ten years, Tajima-mori returned with the fruit. Unfortunately, the Emperor had died before he returned. Tajima-mori, deeply and keenly grieving and lamenting before the tomb of the Emperor, died of sorrow.[18]

The episode of Tajima-mori in the *Kiki* is reminiscent of that of Hsü Fu in the two respects; one, the Emperor wished to get the medicine or the fruits of eternal life; two, the Emperor's wish was not filfilled. The fruit in the land of "Tokoyo," "tokijiku-no-kaku-no-mi," must be similar to the elixir of life in Hsü Fu's story.[19] An author's note in the *Kiki* explains that "tokijiku-no-kaku-no-mi" is today's 'citrus tachibana.' From the context, the fruit is possibly a miraculous one with supernatural power. Because of the legend of Tajima-mori and his fruit, he came to be regarded as the patron god of Japanese confectionery.

In Japan the concept of the "Tokoyo," the eternal land of eternal peace and fertility, is found from ancient times, but the concept of the land where the immortals live in eternal youth is originally a Chinese one. Some scholars suggest that the ideas of "Tokoyo" and "tokijiku-no-kakku-no-mi" in the tale of

---

[17] Kanaya (1973) 294-301, esp. 296; Cf. Mekada, Soji (1969) 334, 343, 481.

[18] Kurano (1958) 202-3; Sakamoto (1967) 279-81, esp. 280-1; Philippi (1968) 226-7; Aston, Nihongi (1972) 186-7; Moriya (1989) 17-8; Camberlain, Kojiki (1981) 243-4.

[19] Cf. Doi, Kodai (1977) 413-4; Obayashi, Tajima-mori (1987) 13-20, esp. 15-6, 20; Miura (1974) 43-52, esp. 44-7; Moriya (1989) 1-22, esp., 10, 18-9.

Tajima-mori are influenced by the Chinese idea of the immortals and eternal youth, although this is not an established theory.[20]

It is true that the idea of the three holy mountains of China, especially the P'eng-lai belief, had an influence on the Japanese way of thinking. It can be seen in the art of garden architecture.[21] When ancient Chinese people built a garden, they made the holy three mountains in a pond shaped like Mt. P'eng-lai, Mt. Fang-chang, and Mt. Ying-chou. In the anthology of the *Monzen* (*Wenxuan*), from the reign of the Emperor Wu-ti of the Han Dynasty, a poem is included that mentions the art of gardening and the three holy mountains in a pond.[22]

The concept of the layout of the garden was then introduced into Japan through the Korean Peninsula.[23] In the case of the layout of a Japanese garden, only one mountain, called "Naka-jima" ('an island in a pond') and symbolising Mt. P'eng-lai, was created in a pond. This style belongs to the main current of Japanese garden practice and can be seen in the "Shinden-zukuri," a typical architectural style of mansion for nobles in the Heian Era, from the eighth to the twelfth century. Mt. P'eng-lai was gradually regarded as identical with Mt. Sumeru of Japanese Buddhism. Mt. Sumeru is said to be the most sacred high mountain lying at the centre of the world.

The spirit of respect for the idea of Mt. P'eng-lai was found in Japan in the planning of Heijo-kyo, today's Nara City, when the Imperial city was constructed in the eighth century. The *Shoku-Nihongi*, the official historical documents of ancient Japan from the seventh century to the eighth century, records the Imperial Edict of Empress Regnant Genmei concerning the transfer of the imperial city to Heijo-kyo as follows:

The topography of Heijo, the Imperial City, fits the principle of the four gods (three beasts and a bird) for the protection of the City. Furthermore the three mountains around the City, (Mt. Kasuga-yama, Mt. Nara-yama, and Mt. Ikoma-yama, or Mt. Unebe-yama, Mt. Kagu-yama, and Mt. Miminashi-yama), guard the City and make it calm and safe.[24]

Hiroyuki Kaneko, the well-known archaeologist, suggests that these three mountains are the symbol of the three holy mountains in China.[25] The topography implies that the Imperial Palace of Heijo-kyo, Nara, mirrors the earthly paradise and so is the ideal residence for the Emperor, the head and master of the Imperial Palace. These documents and the historical buildings

---

[20] Shimode (1986) 58-61, esp. 60.

[21] Miyamoto (1998)12-15, 36-7,44-5; Kaneko, Teien (2002) 46-52.

[22] Obi (1983) 82-4, 135-6.

[23] Kobayashi (1987) 1-32, esp. 16-24; Kaneko, Teien (2002) 46-54, esp. 46-9.

[24] Trans. by the writer. Cf. Kodama (1960) 151-2; Naoki (1986) 100; Aoki (1989) 131.

[25] Kaneko, Bunkazai (1983) 35-42, esp. 37.

prove that the spirit of respect for the three holy mountains, especially for Mt. P'eng-lai, was readily adopted in Japan.

The *Nihon-shoki* contains the official historical documents, and the tale of Tajima-mori is recorded there. This does not mean that the tale is to be regarded as historical fact. The meaning of "history" in those days in Japan resembles that of the medieval European sense of "history," in which not only the historical facts but also myths and legends were included.

IV

Classical Japanese literature, apart from the historical documents, received creative inspiration from the belief in Mt. P'eng-lai. It was naturally influenced by the idea of the eternal land. Not just a few narratives of this motif were composed and recited from the eighth century to the sixteenth century, for example, the tale of Urashima and the tale of Kaguya-hime, the *Tale of Bamboo-Cutter*.[26] Both tales are very famous in Japan, have been handed down from generation to generation, and have had a great influence on Japanese literature.

As is usual in Japanese culture, Japanese literature also had a development of its own that is especially characteristic of Japan. Chinese people were generally fond of a strange and marvellous tales and stories, while, in Japan, as early as the eighth century, an early sign of realism--that may lead to the *Tale of Genji*--is seen in the narratives.

1. The Tale of Urashima

The Tale of Urashima was recorded in the *Nihon-shoki*, in the *Tango-no-kuni Hudoki* (the local history of the Tango-no-kuni district, today's northern part of Kyoto), in the *Man'yo-shu*, and in the *Otogi-zoshi* (books of illustrated stories published from the fourteenth to the sixteenth centuries).[27] It is a typical legend of a man who marries a deity and travels to and returns from another world.

The tale of Urashima is first recorded in the *Nihon-shoki* as factual, having taken place in the reign of the Emperor Yuryaku, who is assumed actually to have existed in many legends.

---

[26] Katagiri (1972) 51-110; Horiuchi (1997) 1-76; Kawabata & Keene (1998) 1-177.

[27] Sakamoto (1967) 497; Aston, Nihongi (1972) 368; Chamberlain, Urashima (1880) 33-6; Aston, Urashima (1904) x-xx, esp. xvii-ix; Ichiko (1958) 337-85; Oshima (1974) 414-24 Mizuno, Hudoki (1975) 210-5; Suga (1991) 103-5.

22nd year,
… Autumn, 7th month. A man of Tsutsukaha in the district of Yosa in the province of Tamba, the child of Urashima [the original text reads Urashima-ko,] of Mizunoye, went fishing in a boat. At length he caught a large tortoise (or turtle), which straightway became changed into a woman. Hereupon Urashima's child fell in love with her, and made her his wife. They went down together into the sea and reached Horai San (Mt. P'eng-lai), where they saw the genii. The story is in another Book. [28]

The tale is generally regarded as a legend or a literary narrative. The story is found also in the *Man'yo-shu*, and in the *Tango Hudoki*. There are several English translations of the Urashima in the *Man'yo-shu*, for example, by Basil Hall Chamberlain in 1880, by W. G. Aston in 1904, and by Teruo Suga in 1991. Here is the rough plot:

Once a youth went on a fishing trip and met a princess of the god of the sea. They went hand in hand to the immortal land. They lived very happily in a gorgeous palace where they would never grow old or die. Thus three years passed. One day the youth happened to remember his parents. Urashima told the princess that he would like to go to see his parents and also that he would return the next day.

The Princess gave him a casket and told him not to open the casket if he wanted to come back to the palace. Urasima started off for his home. When Urashima arrived at the village where his parents lived, the village was all different and quite unfamiliar to him. He could not find his house. He became very uneasy and anxious. It turned out that the three years at the Palace under the sea equated with three hundred years in his world.

It occurred to him that he could find his house if he opened the casket. As he opened it, a white cloud came forth from the casket, and Urashima was very surprised. All of sudden, he became an old man. His body decayed and wrinkled, and his black hair grew white. His breath grew fainter and at last he fell down and died.

In the *Man'yo-shu*, a tortoise does not appear in the story but Urashima did meet the daughter of the god of the sea.[29] In the *Tango Hudoki*, Urashima caught a tortoise which shone in five colours. When he put it in the boat and slept, the tortoise changed into a beautiful lady. She was the Princess of Mt. P'eng-lai, and they went to the island of Mt. P'eng-lai, which is surrounded by the sea on all sides. When three years had passed he yearned for home. In this case what

---

[28] Trans. by Aston, Urashima (1998) 368. Aston reads Urashima-ko as Urashima-no-ko, that is 'the child of Urashima'. The meaning of '-ko' in those days must be the suffix of the name of a man, not 'child'. Cf. Mizuno, Hudoki (1975) 31-49; Obayshi, Shinwa (1986) 76-8.

[29] Kojima, Nihon-shoki (1996) 206-7; Uegaki (1997) 473-79; Kojima, Man'yo-shu (1995) 414-7; Chamberlain, Urashima (1880) 33-5; Aston, Urashima, (1904) x-xx, esp. xvii-ix; Suga (1991) 103-5; Sakamoto (1967) 497; Aston, Nihongi (1972) 368; Mizuno, Hudoki (1975) 210-5.

soared up into the sky was not merely a white cloud but also a sweet fragrance with particles.[30]

The point of the tale is that Urashima went to the ideal earthly paradise and became an immortal man, and yet he lost what he got there--everlasting happy life--because of his own desire to go back to his home and because of his violation of the taboo. This type of tale is not unique to Japan, and similar tales can be found the whole world over. It usually appears as a kind of Rip Van Winkle tale, but from the viewpoint of thematic elements, this tale resembles most closely the legend of Oisìn in Ireland.[31]

Dr. Sakuwo Mekada, referring to a version of the tale, "Mizunoe Urashima-ko," from the end of the seventh century, which is said to be the original story of the tale of the *Tango-no-kuni Hudoki*, pointed out that the tale may not be mere folklore, but possibly a composition with literary intention.[32] She suggests that the narrator composed the tale of Urashima-ko, basing it on the oral folklore in the Tango District and influenced by the tale of Hei-wang-mu, the Queen Mother of the West in China. Hei-wang-mu is an immortal woman. Once a king of Zhou went to Kunlun in the western part of China and met her. The king is said to have forgotten to go back to his palace. Another legend tells that when the Emperor Wu-ti of the Han Dinasty desired and requested a long life, Hei-wang-mu came down from heaven and gave him seven peaches of a special kind. From Dr. Mekada's point of view, the tale of Urashima-ko is strongly influenced by Chinese Taoist teachings on immortality. According to her, although the tale of the Urashim was composed as early as the seventh century, some attempt at fiction can be found in the tale. Judging from the dramatic ending of the tale, her view is reasonable.

In any case, as with the other tales of this kind, the moral of the immortality story is, if it has one, that an ordinary man of this world cannot stay forever in paradise, but must return to his own world. In this sense, the tale of Urashima also implies that the ideal earthly paradise P'eng-lai is Utopia, i.e., Never-Never-Land.

---

[30] Shimode (1986) 204. He suggests the fragrance is the everlasting life that Urashima lost.

[31] Doi, Shinwa (1973) 19-25.

[32] Mekada, Urashima-ko (1960) 2-11.

2. Kaguya-hime in the Tale of the Bamboo-Cutter

Another tale is the *Taketori-monogatari* (*The Tale of the Bamboo-Cutter*), one of the most beautiful Japanese stories of the ninth century.[33] A summary of the story reads as follows:

Once upon a time, there lived an old couple: Okina, a bamboo cutter, and Ouna, his wife. One day Okina found a shining bamboo stalk. When he cut it down, he found inside it a very pretty baby girl. They brought her up with the greatest care and love.

The girl, named Nayotake-no-Kaguya-hime, grew very beautiful so quickly as to become an adult in only three months. Hearing of beautiful Kaguya-hime, many princes came to Okina's mansion to court her. Even the Emperor wished her to serve him at Dairi, his Palace. Kaguya-hime would not accept any of them. Nevertheless they kept on courting her. At last, owing to their earnest desire, she gave in and agreed to see the wooers on the condition that they could succeed in solving a problem. Five wooers wanted to face the challenge.

In short, all of them failed. About three years passed without any particular strife; however, Kaguya-hime began to be absorbed in contemplation, looking up at the moon. She wept bitterly as she gazed up at the mid-autumn full moon. [The mid-autumn full moon has been loved as the most beautiful full moon in the entire year, and the Japanese love, even now, a ceremony for the moon.]

Okina and Ouna wondered and worried terribly, and asked why she was weeping so bitterly. At last, Kaguya-hime confided in them. To tell the truth, she was not a mortal woman of this world, but was of the moon, the other world. She had lived in this world to purify herself of a sin committed in the world of the moon. As her sin had been absolved, it was time for her to go back to the moon.

Okina and Ouna, awfully grieved, at last implored the Emperor to help them. The Emperor, strongly shocked, sent Samurai warriors to protect Kaguya-hime. About the hour of "Ne-no-koku," around midnight, the people of the moon came down from the sky to take her away. In spite of the great many soldiers and archers, the people in the moon easily opened the door.

Kaguya-hime handed over to a vassal of the Emperor a letter to the Emperor with a portion of the medicine of everlasting life. Then she donned a "Hagoromo," a gown for the people of the moon, and soared up into the sky to be with the other moon people.

The Emperor, hearing the whole story, asked what place was nearest to the sky. A vassal answered that it was the mountain in the district of Suruga, today's Shizuoka Prefecture. Then the Emperor ordered that the medicine be brought to the top of the mountain and burnt. Since then the mountain has been called "Fushi-no-yama," the mountain of everlasting life. From there a stream of smoke climbed up into the sky. Today the stream of smoke has stopped soaring, but even now it is called the mountain of everlasting life, that is, Mt. Fuji.

---

[33] Horiuchi (1997) 1-76. Kawabata & Keene (1998) 1-177.

As Prof. Sukeyuki Miura has pointed out, the story of Tajima-mori is similar to that of Kuramochi-no-miko, a character in the tale of Kaguya-hime.[34] A prince, wishing to face her challenge, had to fetch her a branch of the tree that grows in the Eastern Sea on the mountain called Mt. P'eng-lai. Kuramochi-no-miko went out to fetch the branch, but soon returned and hid himself for many months and had the metalsmiths make a branch like the one Kaguya-hime asked for. The branch was so wonderfully and subtly made that Kaguya-hime at first believed it a true branch of Mt. P'eng-lai. However, the metalsmiths suddenly entered Kaguya-hime's mansion just as Kuramochi-no-miko was telling of his journey full of hardships, and asked him to give a reward for the branch. Thus his fakery was known.

Prof. Sadakazu Fujii also suggests a relationship between the tales of Kaguya-hime and of Tajima-mori.[35] In Japanese traditional pronunciation, the difference in the articulation between a voiceless stop and a voiced stop is not very distinct, and, in many cases, especially in ancient times, the two stops are regarded as the same sound. Thus, Kaguya-hime can be pronounced Kakuya-hime. In this case, the sound "kaku" is found both in Kaguya-hime and in "Tokijiku-no-kaku-no-mi" of Tajima-mori.

There should then be a common nature to the two tales. As the meaning of "kaku" is 'fragrant,' 'sweet-smelling,' and 'aromatic,' Kaguya-hime, naturally radiant with beauty, is thought to be fragrantly beautiful. Therefore, according to Prof. Fuji, the fruit that Tajima-mori brought back should also be exquisitely sweet, delicious, and fragrant, just as was Kaguya-hime. As the tale narrates, she was so exquisitely sweet and fragrant as well as radiantly beautiful that she was named Kaguya-hime; "kaguya" means 'fragrant and radiant.'

Although the fruit of Tajima-mori is said to be a kind of citrus, from ancient times the peach has been a special kind of fruit with a magical divine power in China, an idea that was introduced into Japan. In the *Kiki*, Izanagi, a male god, escaping from his wife Izanami's running after him, had a narrow escape owing to a peach.[36] In various folktales and nursery tales also told in Japan, a similar kind of story is found. Accordingly the fruit of Tajima-mori is, insofar as it was the miraculous fruit of eternal life, equivalent to the peach of Hsu-wang-mu.[37]

---

[34] Miura (1974) 43-52.

[35] Fujii, "Kaguya-tanjo" (1985) 54-60, 58-9; Fujii, Taketori (1991) 23.

[36] Kurano, Kojiki (1958) 66-7; Phillipi (1968) 65; Chamberlain, Kojiki (1981) 40.

[37] Cf. Ando (1981) 1-10; Sone, "Kaguya-shokuzai" (1985) 62-9; Sone, "Kaguya-Okina" (1987) 18-33.

102

V

The above-mentioned tales, the tale of Sukuna-hikona, the tale of Hsü Fu, the tale of Tajimamori, the tale of Urashima, and the tale of Kaguya-hime, all tell of people of this world who went to another world, or vice versa. In all cases, the tales imply that another world lies far separated from this world. That world is never easily reached.

It is noteworthy that in the tale of Kaguya-hime P'eng-lai is located in the Eastern Sea. This idea contradicts the traditional Japanese idea that the island of Mt. P'eng-lai should lie to the west of Japan. Probably the words "in the Eastern Sea" in China influenced the tale.

The Chinese, on their very large continent, envisioned Mt. P'eng-lai in the Eastern Sea. As Japan is in the eastern sea of China, some Chinese thought Mt. P'eng-lai was in Japan, whereas the Japanese thought "Tokoyo" to be in China or on the Korean Peninsula.[38] Both ancient Japanese and Chinese very much longed for such a world. At the same time they perceived that the world of the immortals could not be attained in this world but had to be located in another world.

One more noteworthy point is that the tales of Kaguya-hime and of Urashima are not merely a myth and a legend. The story of Kaguya-hime is a creative narrative story of the ninth century, and the tale of Urashima, although an early version is recorded in the *Nihon-shoki*, is almost entirely a creative narrative story.

In the three stories of Sukuna-hikona, of Hsü Fu, and of Tajima-mori, P'eng-lai or "Tokoyo" is depicted as an actual place, whereas the story of Kaguya-hime tells of it in a somewhat different way. In the last scene of Kaguya-hime, the Emperor says that the medicine is to be burnt on the top of the highest mountain in Japan. It implies that the narrator of the story believed that immortality does not pertain to the world of mortals. The last scene is a symbolic description that suggests the narrator's own original idea. The narrator, though restrainedly, breaks away from the impossible dream of the eternal life of the immortal people. This consciousness differs from that of the traditional mythical narrators. Thus the story of Kaguya-hime comes at the point where Japanese narratives depart from myths and legends of traditional style in ancient times.

This is also the case of the tale of Urashima. The last scene, in which the protagonist at once changed from a youth to a very old man and died, has a dramatic effect and may give the audience a keen shock. His sudden change and subsequent death suggest the impossibility of mortals' primordial desire for eternal life at the same time. Although the two tales themselves appear as sheer fancy to most modern realists, and even though the step may be only a slight one,

---

[38] Verschuer (1999) 4, 151-70.

the tales of Kaguya-hime and of Urashima may be the first step to realistic novels in modern times.

It does not mean that the story of Sukuna-hikona, that of Hsü Fu, and that of Tajima-mori are drawn completely from the imagination and have no basis in reality. They are, as it were, myths recorded in the historical documents. However, myths are often an allegory of reality; myths hide a truth in themselves, i.e., all these stories express an aspect of reality in their own way according to their genre.

Ancient people perceived the true meaning of these tales. This is why they believed and accepted the mythical implication of the eternal land with the immortal people with an elixir of life. However, at the same time, in real life they also accepted mortal life as it is: the mortal comes to an end of life in this world. In the tales of Urashima and of Kaguya-hime, the perception was described in a way different from that of the mythical tales.

In the case of Urashima, a myth was reborn as a literary tale and the tale itself came to be changed. A later version more familiar than those of the *Man'yo-shu* and of the *Tango Hudoki* tells another story.[39] Urashima rescued a tortoise from the naughty village children, who were roughly playing with it. As a reward Oto-hime, the princess of the Ryugu-jo, the Dragon Palace under the sea, sent a messenger to fetch him back to the Ryugu-jo. The tortoise was a family member of Princess Oto-hime. Urashima, forgetting everything of his world, spent many days there in pleasure and delight. The final scene, in which he immediately turns into an old, old man, is the same. The dramatic impact of the final scene never changed in all the later variations.

Urashima usually is said to have gone to the palace under the sea, not to the island of Mt. P'eng-lai. In the case of Kaguya-hime, she went to the moon. If the meaning of the "island" is defined as the place separated from the mainland, the palace under the sea, the Ryugu-jo, as well as the moon are also kinds of islands.

In both tales, the image of another world is never changed. The land cannot be reached by mortal men of this world, but the narrators gave up the image of Mt. P'eng-lai. Both the narrators made the concept of the eternal land wide and spacious. They invented a new eternal land, the moon and the Ryugu-jo.

In the case of the tale of Kaguya-hime, the concept of the island of Mt. P'eng-lai persists as in the episode of Prince Kuramochi-no-miko. However, here the world of the eternal land is the moon in the sky. It gives an image of an island surrounded on all sides by the sea as the sky is often compared to the sea. From this viewpoint the moon of Kaguya-hime is also the island in the sea. In the case of the tale of Urashima, the idea of the island itself is forsaken. His palace is under the sea.

---

[39] Mizuno, Kodai (1975) 3-7.

In both cases the image of another world is just one of the most appropriate images born in Japan, which itself consists of islands, surrounded by the sea on all sides. The impossible dream of another world as well as of eternal life is universal. However, in ancient Japan, the desire for a far distant land over the sea must have been very strong and keen. Living without going out of the regions they lived in, surrounded by the sea, they might have longed for another life, another world. At the same time as they were living surrounded by the sea, the world they imagined and dreamt might also be an island surrounded by the sea. It is one of the reasons that the other world of both Kaguyahime and Urashima is an island surrounded by the sea. The topographical environment of Japan may have helped the narrators to break new ground for another world.

Bibliography

I. Sources

Abe, Akio, et. al. eds. *Genji-monogatari, 2.* (*The Tale of Genji, Vol. 2.*) Japanese edition. Tokyo: Shogakukan, 1972.

Aoki, Kazuo, et. al. eds. *Shoku-nihongi 1.* (*A Sequel to the Nihongi.* Vol.1) 5 vols. Japanese edition. Tokyo: Iwanami-shoten, 1989.

Aston, W. G., trans. *The Nihongi: Chronicles of Japan from the Earliest Times to A. D. 697.* London: the Japan Society, 1896; Rutland, Vermont and Tokyo: Tuttle, 1972.

--------. "The Legend of Urashima." *A Grammar of the Japanese Written Language.* Third edition. London: Luzac & Co.; Yokohama: Lane, Crawford & Co., 1904.

Chamberlain, Basil Hall, trans. "Fisher Boy Urasima" *The Classical Poetry of the Japanese.* London: Trübner & Co., 1880.

--------. trans. *The Kojiki: Records of Ancient Matters.* Original first printing Yokohama: Lane, Crawford, 1883; Rutland, Vermont & Tokyo, Japan: Charles E. Tuttle Company, 1981.

Honda, Wataru, et. al. eds. and trans. *Chugoku Koten Bungaku Taikei, Dai 8 kan, Hobokushi, Retsusenden, Shinsenden, Sengaikyo.* (*An Anthology of Classical Chinese Literature, Vol. 8,*) Japanese edition. Tokyo: Heibon-sha, 1969.

Horiuchi, Hideaki, et. al. eds. *Taketori-monogatari, Ise-monogatari. Shin Nihonn Koten Bungaku Taikei, 17.* (*The Tale of the Bamboo-Cutter, Tale of Ise. New Edition of an Anthology of Classical Japanese Literature, 17.*) Japanese edition. Tokyo: Iwanami-shoten, 1997.

Ichiko, Teiji, ed. *Urashima Taro.Otogi-zoshi. Nihon Koten Bungaku Taikei 38.* (*The Tale of Urashima. Books of Illustrated Stories. An Anthology of Classical Japanese Literature.*) Japanese edition. Tokyo: Iwanami-shoten, 1958.

Kanaya, Osamu, et. al. trans. "Toumon-hen." *Ressi; Roshi, Soshi, Gosi. Chugoku Koten Bungaku Taikei, Vol. 4.* ("Tang-wang wen bian." *Lao-zi, Zhuang-zi, li-e-zi, and wu-zi. An Anthology of Classical Chinese Literature. Vol. 4.*) 60 vols. Japanese edition. Tokyo: Heibon-sha, 1973: 1987.

Katagiri, Yoichi, ed. and tr. *Taketori-monogatari; Ise-monogatari Yamato-monogatari, Heiju-monogatari. Nihon Koten Bungaku Zenshu 8.* (*The Tale of the Bamboo-Cutter, The Tale of Ise, The Tale of Yamato, The Tale of Heiju. An Anthology of Classical Japanese Literature, 8.*) Japanese edition Tokyo: Shogakukan, 1972.

Kawabata, Yasunari, *The Tale of the Bamboo Cutter.* Trans. Donald Keene. Tokyo: Kodan-sha International, 1998.

Kojima, Noriyuki, et. al. eds. and trans. *Man'yo-shu.* (*A Collection for a Myriad of Classical Japanese Poems, 2.*) 4 vols. Japanese edition. Tokyo: Shogakukan, Vol. 1. 1971; Vol. 2. 1972; Vol. 4. 1975.

--------. *Nihon-shoki, 2.* (*The Nihon-shoki, No. 2.*) 3 vols. Japanese edition. Tokyo: Shogakukan, 1996; 2004.

Kurano, Kenji, ed. *Kojiki. Nihon Koten-bungaku Taikei 1, Kojiki, Norito.* (*An Anthology of Classical Japanese Literature 1, the Kojiki, the Norito.*) eds. by Kenji Kurano and Yukichi Takeda. Japanese edition. Tokyo: Iwanami-shoten, 1958.

Mekada, Makoto, ed. "Tenmon." *Soji* ("Tian- wen" *Chu-ci*). *Shikyo Soji. Chugoku Koten Bungaku Taikei. Vol. 15.* (*The Shih Ching and Chu-ci; The Book of Odes and Songs; Words of Chu. An Anthology of Classical Chinese Literature. Vol. 15.*) 60 Vols. Japanese edition. Tokyo: Heibon-sha, 1969.

Mizuno, Yu "Tango-no-kuni Hudoki Itsu-bun. In'yo-shiryo Genbun Shusei". *Kodai Shakai to Urashima Densetsu,* ("Quotations from Hudoki of Tango Prefecture: A Collection of the Quotations from Historical Materials". (*Ancient Society and Urashima Legends*) Vol. 1 in two volumes. Japanese edition. Tokyo; Yuzankaku, 1975, 210-5

Motoori, Norinaga, *Kojikiden, Zoho Motoori Norinaga Zenshu, Dai 3,* (*An Annotated Kojiki. The Enlarged Version of the Complete Works of Motoori Norinaga, Vol. 3.*) Revised by Toyokai Motoori. Japanese edition. Tokyo: Yoshikawa-kobunkan, 1902; 1926.

Naoki, Kojiro, ed. and trans. *Shoku-nihongi Vol. 1.* (*A Sequel to the Nihongi Vol. 1.*) Japanese edition. 4 vols. Tokyo: Heibon-sha, 1986.

Noguchi, Sadao, et. al. trans. *Shiba-sen Shiki. Chugoku Koten-bungaku Taikei 10.* (*Ssu-ma Ch'ien's Shih-chi. An Anthology of Classical Chinese Literature 10.*) Japanese edition. Tokyo: Heibon-sha, Vol. 1 1968; Vol. 3 1971.

Obi, Koichi, ed. *Monzen (Wenxuan) 1. Zenshaku Kanbun Taikei 26.* (*Monzen 1, An Anthology of Chinese Literature 26*) Japanese edition. Tokyo: Shuei-sha, 1974; 1983.

Ogawa, Tamaki, et. al. trs. *Shiki-retsuden.* (*Shi ji: Biographies*) 5 vols. Japanese edition. Tokyo: Iwanami-shoten, 1975.

Oshima, Takehiko, ed. and tr. *Urashima Taro. Otogi-zoshi Shu. Nihon Koten Bungaku Zenshu 36.* (*The Tale of Urashima. Books of Illustrated Stories. An Anthology of Classical Japanese Literature 36.*) Japanese edition. Tokyo: Shogaku-kan, 1974.

Philippi, Donald, L., trans. *Kojiki.* Tokyo; University of Tokyo Press, 1968.

Sakamoto, Taro, *Nihon-shoki, 1. Nihon Koten Bungaku Taikei 67.* (*The Nihon-shoki, Part 1. An Anthology of Classical Japanese Literature 67.*) 2 vols. Japanese edition. Tokyo: Iwanami-shoten, 1967.

Saku, Misao, trans. and annotation. *Zoku Kokuyaku Kanbun Taikei. Hakurakuten Zen-shishu. Dai 1 kan.* (*A Sequel to An Anthology of Japanese Translation from Chinese Literature in The Complete Poetical Works of Pai Chiui. Vol. 1.*) Japanese edition. Tokyo: Nihon-tosho-centre, 1978; 1990.

Ssu-ma Ch'ien. *Shi-chi.* (*History of Ancient China.*) 10vols. Peking: Chung-hua Shu-chu, 1959; 1982.

Suga, Teruo, trans. "1704 Poem on the lad of Urasima of Mizunoe; with a poemette". *The Man'yo-shu.* Tokyo: Kanda Institute of Foreign Languages, Vol.2 in three volumes. 1991.

Takagi, Ichinosuke, et. al. eds. *Man'yo-shu.* (*A Collection for a Myriad of Classical Japanese Poems.*) 4 vols. Japanese edition. Tokyo: Iwanamishoten, Vol.1.1957, Vol.2. 1959, Vol.4. 1962.

Takigawa, Kametaro, ed. *Shiki Kaichu Kosho.* (*An Annotated Shih-chi.*) 10 vols. Japanese edition. Tokyo: Toho-bunka-gakuin-Tokyo-Kenkyujo, Vol. 1-2, 4, 1956; Vols.3-6, 1956-7; Vol. 9, 1959.

Tyler, Royall, trans. *The Tale of Genji.* New York, *Penguin Books*, 2003.

Tsai-fa, Chen, William H. Nienhauser, Jr. and Robert Reynolds, trans. *The Grand Scribe's Records, Vol. 1,* Ed. by William H. Nienhauser, Jr. *The Basic Annals of Pre-Han China by Ssu-ma Chi'ien.* Bloomington and Indianapolis: Indiana Univ. Press, 1934.

Uegaki, Setsuya, ed. and trans. *Hudoki. Shinpen Nihon Koten Bungaku Zenshu 5.* (*A Local History. A New Edition of An Anthology of Classical Japanese Literature. No. 5.*) Japanese edition. Tokyo: Shogaku-kan, 1997; 2003.

Waley, Arthur, trans. *A Wreath of Cloud, the third part of "The Tale of the Genji"* London: George Allen & Unwin Ltd. first published in 1927.

108

Wilkinson, Endymion. *Chinese History A Manual*. Revised and enlarged edition. Cambridge (Massachusetts) and London: The Harvard University Asia Center for the Harverd-Yenching Institute and distributed by Harvard University Press, 2000.

II. Secondary Literature

An Shibin, et. al., *Jofuku Densetsu wo Saguru* (*A Study of the Legends of Hsü Fu*) Japanese edition. Tokyo: Shogakukan, 1990

Ando, Shigekazu. "Kaguya-hime no Shokuzai-tan – Taketori-monogatari wo Kanryu surumono –." *Kokugo-kokubungaku-ho*. ("Story of Purification of Kaguya-hime – A Stream Flowing through the Tale of the Bamboo Cutter." *The Essays of Japanese Language and Literature*.) Japanese edition. No. 38. Kariya: Aich Kyoiku Daigaku Kokugo Kokubungaku Kenkyushitsu, 1981.

Asakura, Haruhiko, et. al. *Shinwa Densetsu Jiten*. (*A Dictionary of Myths and Legends*.) Japanese edition. Tokyo: Tokyodo-shuppan, 1997.

Cheng Tianliang, *Jofuku Kirino kanatahe*. (*Jofuku Over the Mist*.) Trans. by Shoji Ikegami. Japanese edition. Tokyo: Daiichi-shobo, 2000.

Doi, Kochi, *Kodai Densetsu to Bungaku. Doi Kochi Chosaku-shu, Dai 2 kan*, (*Ancient Legends and Literature. The Selected Essays of Doi Kochi, Vol. 2*,) Japanese edition. Tokyo: Iwanami-shoten, 1977.

--------. *Shinwa. Densetsu no Kenkyu*. (*A Study of Myths and Legends*.) Tokyo: Iwanami-shoten, 1973.

Fujii, Sadakazu. "Kaguya-hime – Taketori-monogatari Shujinko no Tanjou." *Kokubun-gaku*. ("Kaguya-hime: The Birth of the Heroin of the Tale of the Bamboo-Cutter." *Japanese Literature*.) Japanese edition. Tokyo: Gakuto-sha, No. 7. 1985. 54-60.

--------, ed. *Taketori monogatari, Yamato monogatari, Utsuho monogatari. Shinpan Koten Arubamu*. (*The Tale of the Bamboo-Cutter, The Tale of Utsuho. A New Edited Album of Classical Literature*.) Japanese edition. Tokyo: Shincho-sha, 1991.

Kaneko, Hiroyuki. *Shinsen-sekai to Jodo to Shima, Kodai-teien no Shiso – Shinsen-sekai heno Shokei*. (*Concept of the Layout of the Garden in Ancient Times; Longing for the World of the Immortals*.) Japanese edition. Tokyo: Kadokawa-shoten, 2002.

--------. "Tojo hakkutsu shi 7, Heijo-kyo ato". *Gekkan Bunkazai.* ("A History of the Excavation of the Imperial City 7: The Site of Heijo-kyo". *Manthly Cultural Treasure.*) Ed. by Bunka-cho bunkiazai hogobu. Japanese edition. Tokyo: Dai'ichi-hoki, 1983.

Kobayashi, Masa'aki, "Horai no Shima to Rokujo-in no Tei'en". *Tsurumi Daigaku Kiyo.* ("Island of Mt. P'eng-lai and the Garden of Rokujo-in." *Bulletin of Tsurumi University.*) 24-I. Japanese edition. Yokohama: Tsurumidaigaku, 1987. 1-32.

Kodama, Kota, ed. *Shiryo ni yoru Nihon no ayumi: Kodai-hen* (*A Japanese History from the Historical documents*). Tokyo: Yoshikawa kobun-kan, 1960.

Mekada, Sakuwo. "Iyobe Umakai saku 'Mizunoe Urashima-ko den' (kasho) ko." *Kokubungaku: Gengo to Bungei.* ("A Study of 'Mizunoe Urasima-ko' (provisional) by Iyobe Umakai". *Japanese Language and Literature.*) Japanese edition. Tokyo: Meiji-shoin, 1960 no. 11.

Mishina, Shoei, *Mishina Shoei Ronbunshu, Vol. 2, Kenkoku-shinwa no sho-mondai.* (*Collected Essays of* Shoei *Mishina, Vol. 2, Various Issues of the Myths of a Country-Foundation.*) Japanese Edition. Tokyo: Heigon-sha, 1971.

Miura, Sukeyuki, "Tajima-mori to Kuramochi-no-miko – Ijigen-kukan no Hiteisei." *Bungaku gogaku.* ("Tajima-mori and Kuramochi-no-miko – Negation of Another Dimention of Space." *Literature, Language.*) Japanese edition. Tokyo: Sanseido, 1974. .

Miyamoto, Kenji. *Zusetsu Nihon Tei'en no Mikata.* (*An Illustrated Guide to Japanese Gardens.*) Japanese edition. Kyoto: Gakugei-shuppan, 1998.

Mizuno, Yu. *Kodai Shakai to Urashima Densetsu.* (*Ancient Society and Urashima Legends.*) Vol. 1 in two volumes. Japanese edition. Tokyo: Yuzankaku, 1975.

Moriya, Toshihiko. "Tajima-mori – Tachibana no Shoraisha." *Shinto-gaku.* ("Tajima-mori – One Who Introduced a Tachibana." *Shinto.*) Japanese edition. No. 142. Japanese edition. Isumo-taisha: Shinto-gakkai, 1989.

Nakamura, Hirotoshi. "Tajima-mori no monogatari" *Kokugo Kokubungaku Ronso.* ("Story of Tajima-mori." *A Collected Essays on Japanese Language and Literature.*) Ed. by the Publishing Association of the Essays in Honour of the Seventieth Birthday of Prof. Zenmaro Ohta. Japanese edition. Japanese edition. Tokyo: Gunsho, 1988.

Obayashi, Taryo. "Tajima-mori to Takenouchi-no-Sukune." *Higashi Ajia no kodai-bunka*. ("Tajima-mori and Takenouchi-no-Sukune." *Ancient Culture of Eastern Asia*.) No. 50. Japanese edition. Tokyo: Daiwa Shobo, 1987, 13-20

--------. *Shinwa no keifu – Nihon Shinwa no Genryuu wo saguru*. (*The Genealogy of Myths: Searching for the Roots of Japanese Myths*.) Japanese edition. Tokyo: Seido-sha, 1986.

Okamoto, Ken-ichi, "Horaisan to Fusoju heno Akogare – Nihon-bunka no Koso no Tankyu. 1." *Journal of Human Cultural Studies*. ("Penglai and Fusang, or the Eastern Mountain and Tree of Life – As Archaeological Study on Symbolism of Utopia in Ancient Japan. Part 1.") Japanese edition. Kyoto: Human Cultures Association of Kyoto Gakuen University, 1999.

Orikuchi, Shinobu. "Haha-ga-kuni he, Tokoyo he," "Kodai Kenkyu" *Orikuchi Shinobu Zenshu. Dai 2 kan*. ("To the Country of Mother, To the Eternal land," "Study on Ancient Times." *The Complete Works of Origuchi Shinobu. Vol. 2*.) Japanese edition. Tokyo: Chuo-koron-sha, 1955;1987.

Shimode, Sekiyo. *Kodai Shinsen Shiso no Kenkyu*. (*A Study of Ancient Taoist Immortality Thought*.) Japanese edition. Tokyo: Yoshikawa-kobunkan, 1986.

Sone, Seiichi, "Kaguya-hime no Koutan to sono Shokuzai" *Heian Bungaku Kenkyu*. ("The Birth of Kaguya-hime and Her Purification". *Study on Japanese Literature in the Heian Era*.) No. 73. Japanese edition. Kyoto: Heian Bungaku Kenkyu-kai, 1985.

--------. "Kaguya-hime to Taketori no Okina – 'Mukashi no Chigiri', 'Kudoku', 'Tsumi' no saikento." *Hanazono Daigaku Bungaku Ronkyu*. ("Kaguya-hime and Bamboo-Cutter: Reconsideration of 'the Pledge of the Olden Times', 'Virtuous Deeds', 'Sin'." *The Collected Essays of Japanese Literature of Hanazono University*.) No. 15. Japanese edition. Kyoyo: Hanazono Daigaku Koku-bungakkai, 1987.

Takimoto, Hiroyuki. *Chugoku Rekishi Jinbutsu Daizuten – Shinwa and Densetsu hen*. (*An Illustrated Dictionary of Chinese Historical Characters: Myths and Legends*.) Japanese edition. Tokyo: Yushi-kan, 2005.

Tanigawa, Ken'ichi, *Tokoyo-ron*. (*A Study of 'Tokoyo'*.) Japanese edition. Tokyo: Kodan-sha, 1989.

Tsuda, Sokichi. *Nihon Koten no Kenkyu, 1. Tsuda Sokichi Zenshu, Dai 1 kan*. (*A Study of Classical Japanese Classics, 1. The Complete Works of Sokichi Tsuda, Vol. 1*.) Japanese edition. Tokyo: Iwanami-shoten, 1963.

Tsuji, Shiho. *Jofukuron Ima wo Ikiru Densetsu.* (*On Hsü Fu: His Legends Past and Present.*) Japanese edition. Tokyo: Shinten-sha, 2004.

Von Verschuer, Charlotte "To So niokeru Nihon Horai-kan to Suigin Yunyu". *Asia Yugaku (Intriguing Asia).* No. 3. Tokyo: Bensei shuppan, 1999. 151-70.

Yoshida, Atsuhiko. "Tajima-mori to Tokijiku-no-konomi" *Hojo to Fushi no Shinwa.* ("Tajima-mori and the Fruits lof Tokijiku-no-konomi" *The Myth of Fertility and Everlasting Life.*) Japanese edition. Tokyo: Seido-sha, 1990.

Asia Rekishi Jiten. (*An Encyclopedia of Asian History.*) 12 vols. Japanese edition. Tokyo: Heibon-sha, 1960; 1984.

Prof. Yuko Tagaya
Kanto Gakuin University
3-22-1 Kamariya-minami, Kanazawa-ku,
236-8502 Yokohama, Japan
tagaya@kanto-gakuin.ac.jp

Kyoto in Myth and Literature

I

It is often said that to tell the history of Kyoto is to tell the history of Japan.[1] Kyoto was chosen as the capital of Japan in 794. The capital was transferred from Kyoto to Tokyo in 1868 (Edo: Fig. 2). Kyoto lasted about twelve hundred years, a little longer than Constantinople of the Eastern Roman Empire. In those years, the most important core nature of the Japanese people was nurtured.

To appreciate the spiritual nature of the Japanese is to value one of the best parts of the Japanese culture. As is universally true in every other society, culture represents the fruits of our people's collective efforts, which have maintained our noble qualities. This paper considers the nature and essence of the culture of Kyoto from the viewpoint of myth and literature, as well as history.

§ 1. Japanese Palace Cities before Heian-kyo

In ancient times, until about the seventh century, whenever a Tenno, the Emperor of Japan, passed away and a new Tenno ascended the throne, the capital was relocated, because the previous one was regarded to have been polluted by the death of the Tenno.[2] In those days, the thought of death as impure and unclean was widespread among the Japanese people. The locations of the capital cities mostly converge in today's Kinki District: Kyoto, Nara, Shiga, and Osaka. However, not all of them, unfortunately, have been confirmed archaeologically beyond doubt. Fig.1 (Map of Ancient Kinki District) shows the ancient capital cities that have been confirmed by historical documents and archaeological method.

According to the chronological table of Japanese history, capital cities previous to Heian-kyo were Fujiwara-kyo (today's Kashihara-shi in Nara) of Jito Tenno, Heijo-kyo (today's Nara) of Genmei Tenno, and Nagaoka-kyo (today's southern part of Kyoto-fu, between Osaka-shi and Kyoto-shi) of Kanmu Tenno.

---

[1] Kyoto no Rekishi to Bunka. (1988) 20.
[2] Kishi (1981) 13-5.

the Kinki District with the trunk roads
Fig.2
Hokuriku-do
Arachi
Tosan-do
San-in-do
now Kyoto
Heian-kyo
Fuwa
Tokai-do
San-yo-do
Osaka
Suzuka
Nankai-do
Edo
Now Tokyo
Kamakura
Kyushu
Hyuga
by Yuko Tagaya

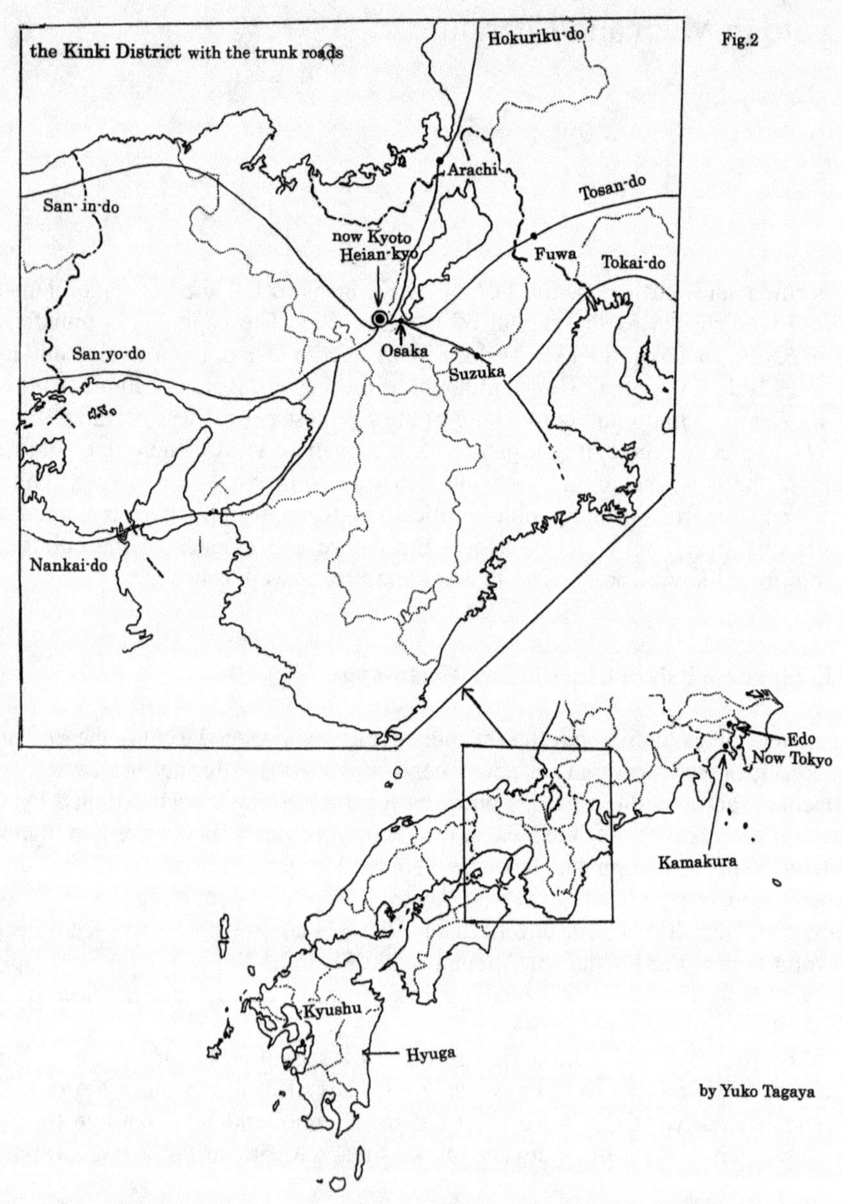

Fig. 2

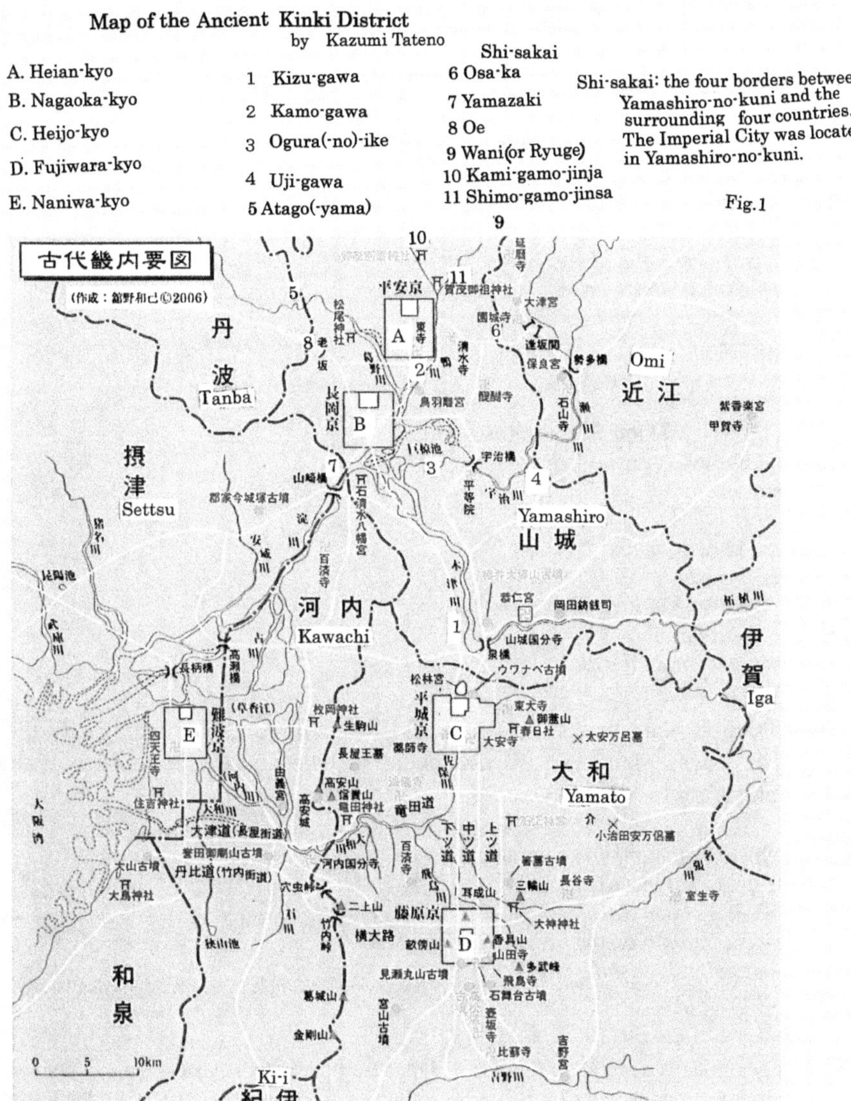

Fig. 1

Both Jito Tenno and Genmei Tenno were Empresses Regnant; the former was the consort of Tenmu Tenno (?-686; r. 673-686) and an Imperial Princess of Tenji Tenno (626-671; r. 667-671). Jito Tenno as well as Tenji Tenno is famous and popular for their *waka* poems, both of which are included in the *Hyakunin isshu*.[3]

The followings are the *waka* poems of Tenji Tenno and of Jito Tenno.

Aki no ta no kariho no iho no toma wo arami
Waga koromode ha tsuyu ni nure tsutsu
Tenji Tenno[4]

Coarse is the rush mat of my harvest-hut,
Sheltering the harvested ears of this autumn day;
So my sleeves are growing wet
With the evening dew.

Haru sugite Natsu kinikerashi shirotahe no
Koromo hosutefu Amano-kagu-yama
Jito Tenno[5]

Lo, spring has gone and
Summer has come,
To Mt. Ama-no-Kagu-yama;
Pure white robes
Are spread to dry in the air.

Heijo-Kyo in Nara was such a beautiful, splendid, and flourishing city that the sight of it was praised in a *waka* poem of Ono-no-Oyu, a courtier in the eighth century:[6]

Awoniyoshi Nara no miyako ha saku hana no
Nihofu ga gotoku ima sakari nari.
In the capital city in Nara, today
The flowers in full blossom are at their height
And beautifully flourishing with
Sweet aromatic fragrance.

---

[3] 'Waka' poem means generally all kinds of Japanese poems, in a narrow sense, Japanese short poems consisting of 5-7-5-7-7 syllables. Hyakunin isshu is a collection of one hundred 'waka' poems, each by a different poet, compiled in the mid-thirteenth century by Fujiwara-no Teika.

[4] Hyakunin isshu, No. 1. Ariyoshi (1983) 16-9.

[5] Hyakunin-isshu, No. 2. Ariyosi, 20-3. Ama-no-Kagu-yama is a mountain in Nara, one of the three important mountains of the Yamato District.

[6] Man-yo-shu, no. 328. Satake (1999) 231. Man-yo-shu is the oldest existing anthology of Japanese poems from the 6th to the 8th centuries. Man-yo means 'ten thousand leaves'.

Kanmu Tenno of the Heian Era ascended to the throne in 781. In 784 he first transferred his Palace City to Nagaoka. Nagaoka-kyo, like Fujiwara-kyo and Heijo-kyo, was a full-fledged Palace City, taking that of China as its model.[7] Because of various circumstances, however, in 794, Kanmu Tenno again transferred his Palace City to Kyoto.

§ 2. Heian-kyo

1) *The Topography of Kyoto*

In ancient times the area of today's Kyoto was a lake, which gradually became a basin, a round valley.[8] Today, Kyoto is topographically a basin land surrounded by mountains and hills on the three sides. To the south it opened onto Ogura(-no)-ike Pond, into which several rivers flowed.[9] Kyoto in this way serves as one of the best regions for a natural fortress.[10]

During the period from about the end of the third century to the seventh century, the Kofun period ('kofun' means a tumulus, an ancient burial mound), some powerful families lived near the lower river bank of the Katsura-gawa River and near Sagano Field.[11]

Kyoto then gradually developed and prospered. According to the *Kiki*, upon seeing the the flourishing of Uji, the southern part of Kyoto, Ojin Tenno, the legendary Emperor of about the beginning of the fifth century, who was on his way to Omi-no-kuni, today's Shiga prefecture, composed a poem.[12]

Chiba no Kadono wo mireba
Momochitaru Yaniwa mo miyu
Kuni no ha mo miyu

Ojin Tenno

---

[7] Ashikaga (1994) 20-3; Nishikawa, Toshi no Shiso (1973) 71-5; 81-91.

[8] Nishikawa, Kyoto (1997) I. 4-6.

[9] Ogura(-no)-ike Pond was once very large but was reclaimed and at last completely filled in 1941.

[10] 'The Imperial edict of Kanmu Tenno' in Kodama (1960) 220; Takinami (1991) 41-2; Fujiwara (1944) 57.

[11] Nishikawa, Kyoto. I. 8-9.

[12] Kurano (1968) 240-3; Sakamoto (1967) 365. Kiki means the Kojiki and the Nihon-shoki. The Kojiki is the oldest existing historical documents compiled in 712 ordered by Genmei Tenno. The Nihon-shoki is the oldest official history of Japan ordered by Tenmu Tenno from the Jindai period, the age of the gods, to the reign of Jito Tenno, compiled in 720.

118

As I look on the Moor of Kadzu in Chiba.
both the hundred thousand-fold abundant
house-places are visible, and the land's
acme is visible.

Tr. by Chamberlain[13]

2) The Early Days of Kyoto and Its Myth

A mythical legend in *Yamashiro-no-kuni Fudoki* relates how Kyoto developed.[14] Fig. 1 shows the places Yamato and Yamashiro, and Fig. 2 shows Hyuga.

Once, Kamotake-tsunomi-no-mikoto, a legendary deity, came down to Hyuga, today's Kyushu, an island in the southern part of Japan, from the heavenly palace, to Yamato, today's Nara, as the guide of Jinmu Tenno, the first legendary emperor. Kamotake-tsunomi-no-mikoto himself lived, at first, at the ridge of Mt. Katsuragi-san, in today's Nara. Thereafter he moved to the Kamo district in Yamashiro-no-kuni, today's Kyoto, and then went down to the Kizu-gawa River. Later still he went up the Kamo-gawa River and at last settled at the foot of the mountain located to the north of Kuga-no-kuni near the upper part of the Kamo-gawa River.

Kamotake-tsunomi-no-mikoto married the daughter of a powerful family there and had two children. One day when his daughter, Tamayori-hime, was playing by a brook, a red arrow floated down. She picked it up and put it at her bedside. Then she conceived and a boy baby was born.

When the child had grown up, Kamotake-tsunomi-no-mikoto told him to bring the cup of sake to whomever the child thought was his father. Then the child devoted the wine cup to heaven and rose up to heaven, breaking through the roof. His father was the god of Thunder. Then Kamotake-tsunomi-no-mikoto, his grandfather, named him Kamowake-ikazuchi-no-mikoto, Prince Thunder.[15]

Thus the enshrined deity of Kami-gamo-jinja, a shrine in Kyoto, is this child, Kamowake-ikazuchi-no-mikoto, while those of Shimo-gamo-jinja, another shrine in Kyoto, are his grandfather god and his mother goddess. Aoi-matsri, one of the most famous and familiar festivals in Kyoto, is held by the two shrines.

---

[13] Chamberlain (1981) 304.

[14] Nishikawa, Kyoto, I. 7. Akimoto (1958) 414-5. Fudoki, compiled by the order of the central government during the Nara Era, are the records of the natural features, culture, and history of a region: the provincial historical documents.

[15] Ikazuchi means 'thunder'.

3) The Concept of the Construction of Kyoto, Heian-kyo

1. The Four-Gods Principle

Kanmu Tenno called his new Imperial City Heian-Kyo, because he hoped for peace, safety, and calm. *Hei* means 'peace' and *An* means 'safety and calm'. Heian-kyo is said to have been constructed on a model based on the same principle as Changan in China. According to recent materials and data, however, Heian-kyo may be an advanced form of the previous cities, Heijo-kyo and Fujiwara-kyo, rather than a mere copy of Changan.[16] Fig. 3 shows the location and the shape of the Heian-kyo.

According to the Chinese concept, first of all, the construction of the capital city should be based upon the Four-Gods principle, i.e., at each of the four sides of the City stands one of the Four Gods and they protect the city together.[17] On the east stands Seiryu, Blue Dragon; on the south, Suzaku, Red Bird, a mythical bird like the phoenix in Europe; on the west, Byakko, White Tiger; and on the north, Genbu, a mythic and fabulous animal whose body is made up of a black turtle and a serpent. The above animals represent the Four Gods because of a time-honoured saying. The heavenly body has twenty-eight stations. When they are divided into four parts, each part comprises seven stations. Thus, the easternmost of the four parts is ruled by Seryu, Blue Dragon; the southern one by Suzaku, Red Bird; the western one by Byakko, White Tiger; and the northern one by Genbu, Black Turtle. The constellations of each part make up the figure of each animal mentioned above.

In the case of Heian-kyo, it is popularly said that the Four Gods related strictly to a topographical viewpoint.[18] On the east, there flows the Kamo-gawa River as running water (Blue Dragon); on the south, Ogura(-no)-ike, a pond (Red Bird); on the west, San-in-do or Sanyo-do, a trunk road (White Tiger); and on the north, Funaoka-yama, a mountain (Black Turtle) (Fig. 3).[19] Thus, Kyoto was designed to reflect the concept of the Four Gods principle. Some scholars point out that the Four Gods principle was introduced from ancient China, but that the idea of the close relation between the Four Gods and the topography mentioned above is peculiar to Japan and not found in China.[20]

---

[16] Kishi, 125-30.
[17] Takikawa (1967) 339; Groot (1892-1910). III. 949-52, especially 949.
[18] Nishikawa, Kyoto. I. 12-3.
[19] Nishikawa, Kyoto. I. 4.
[20] Mesaki (1998) 170-75.

2. The Imperial City of Heian-kyo

Heian-kyo was as large as 4.7 kilometers from east to west and 5.2 kilometers from north to south.[21] It was a symmetrical city. The Palace is located on the north of the City. The central street is the Suzaku-oji, 70 meters or more in width.[22] Along both sides of the Suzaku-oji, white walls were constructed. Heian-kyo was so elegantly beautiful and was praised in *wakas*, for example, one by Ise-no-Taifu, a woman poet of the tenth century and one by Sosei-hoshi, a Buddhist priest poet in the tenth century:

> Inishihe no Nara no miyako no yahezakura
> Kefu kokonohe ni nihoi nuru kana
>
> Ise-no-Taifu[23]

> *Double cherry blossoms,*
> *Once being full bloom beautifully in Nara,*
> *Ancient Capital City,*
> *Have today more beautifully and more gloriously*
> *Come into full bloom,*
> *Perfuming this Palace of Miyako with fragrance.*

> Miwataseba yanagi sakura wo kokimazete
> Miyako-zo haru-no nishiki narikeru
>
> Sosei-Hoshi[24]

> *Looking around, all is woven*
> *So amazingly, so richly with*
> *Willows and cherry blossoms,*
> *Miyako of Heian-kyo is now*
> *At the height of Spring.*

4) The Rajo-mon Gate and the Protection of the City

Heian-kyo was located at the end of the trunk roads which spread out in all directions through Japan (see. Fig. 2), just as "All roads led to Rome." Although Heian-kyo was constructed on the model of Changan, and the Rajo-mon, now lost, was constructed as the front gate to the south, no town walls were built

---

[21] Moriya (1994) 6. The opinion as to the size of the City is divided: 4.5×5.2 km in Rekishi-chizu-hon (2006)10, 13; 4.5×5.3 km in Ashikaga (1994) 32.

[22] Nishikawa, Kyoto. I. 4; 84 m in Rekishi-chizu-hon, 13; 85 m in Ashikaga, 32.

[23] Hyakunin isshu, No. 61. Ariyoshi, 256-59.

[24] Kokin Waka-shu, No. 56. Kojima (1989) 34.

around the city. As a result, the border between the inside and the outside of the city was not so clear.

In the present day, the front gate is generally called Rasho-mon, but in those days, it was called Rasei-mon or Raisei-mon.[25] Gradually, it came to be called Rasho-mon, for example in Yo-kyoku, the vocal music of *Noh* drama, an art of Japanese public entertainment. Today, however, in the historical field, it is usually called Rajo-mon.

From the gate of the Rajo-mon, a road stretched to a port, Toba-minato, the Port of Toba. The post stations were also set along the trunk roads with three barrier stations, *Seki* in Japanese: Fuwa-no-seki in Mino, today's Gifu Prefecture; Suzuka-no-seki in Ise, today's Mie Prefecture; and Arachi-no-seki in Tsuruga, today's Fukui prefecture.[26] These barrier stations were, however, abolished after a short time because they caused traffic congestion. The flow of traffic had a higher priority than the defense of the capital. This might have been possible because they had not very many strong foreign enemies in those days.

The Rajo-mon Gate was so magnificent and marvelous, seven *kens* (one *ken* is about 1.8 metres) in width, and two *kens* from front to back. The walls were white, and the pillars were vermillion, with genuine tiles on the roof. On both ends of the roof edge was set an ornamental ridge-end tile. Moreover on top of the roof stood Tobatsu Bishamonten, a god who had the divine power to repel outside enemies.[27]

On both sides of the Rajo-mon, there stood a temple: the To-ji, the Eastern Temple; and the Sai-ji, the Western Temple. Both temples were built to keep the city tranquil and peaceful. Kukai, the head Buddhist priest of the To-ji, started a school, Shugei-shuchi-in, for the common people in 828. He had been deeply impressed by the educational system in Changan when he had studied there.[28] Unfortunately, the school was closed after his death (835). However, his school had much influence on Japanese education for the common people.

---

[25] Ashikaga, 28.

[26] Nishikawa, Kyoto. I. 56; Nishikawa, Toshi. 97-99.

[27] Ishida. 1997, 20-1; Nakamura (1988) 494-5. Nakamura, Compact edition (1981) 1133-4. Tobatsu-bishamonten is a god of Bishamonten (Vaiśravana). According to Buddhist teachings, he is a god of treasure and war, usually clad in armour, with a spear and a treasure house. He has the supernatural power to drive out foreign enemies.

[28] Kanaoka (1979) 117-8; Takahashi (1984) 43-53; Yajima (1984) 245-59.

Heian-kyo

1. the Rajo-mon
2. the To-ji
3. the Sai-ji
4. Rokudo-chinko-ji

Funeral Places

5. Toribeno
6. Sai-no-kawara
7. Adashsino
8. Rendaino

9. Fukakusa
10. Takeda
11. Kamo-gawa no kawara

Fig.3

the Seimei-jinja
the Modori-bashi

North: Genbu,
Black Turtle
(Funaoka-yama)

8

Odoi
(Funaoka-yama)

Heian-kyo

7

West: Byakko,
White Tiger
(San-in-do or San-yo-do)

11

East: Seiryu,
Blue Dragon
(Kamo-gawa)

Suzaku-
oji

4

Shi-sumi of the Impeial Palace

5

3

2

1

(Kamo-gawa)

(San-in-do or San-yo-do)

South: Suzaku,
Red Bird
(Ogura(-no)-ike)

6

9

10

(Ogura(-no)-ike)

by Yuko Tagaya

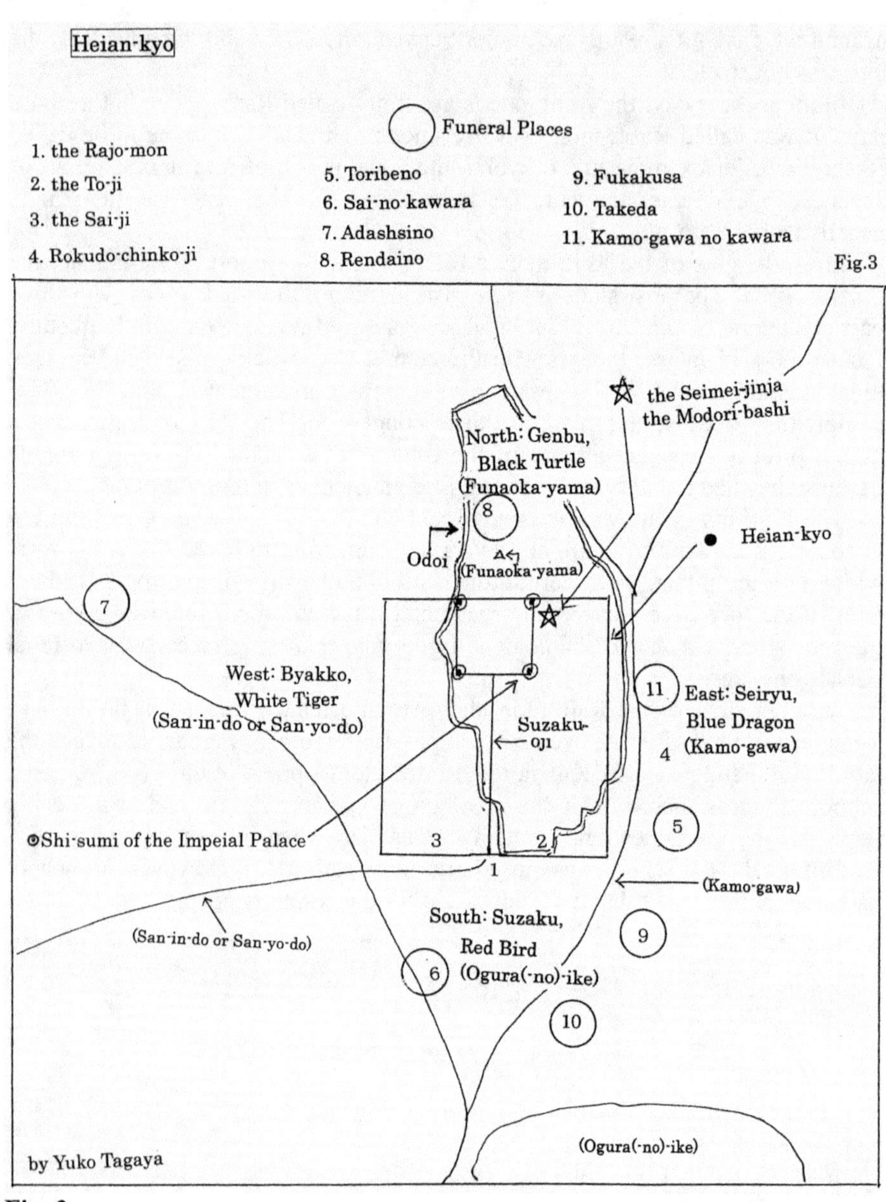

Fig. 3

§ 3. The Borders of the City

1) The Shi-sumi-Shi-sakai-no-matsuri and the Rites[29]

The people of the Heian-kyo did not make much of either the barrier stations or the town walls. It suggests that they seem not to have worried more about living foreign enemies than the ancient Chinese and European people did; rather they seem to have been much more greatly afraid of impurity, wicked spirits, and the curses of the dead. This is why the temples and the shrines were built to enclose the Imperial City. The townspeople held various kinds of religious observances for their own protection against these evil things.

One observance is the Shi-sumi-Shi-sakai-no-matsuri. *Shi* means 'four', *Sumi* means 'a corner point', and *sakai* is 'a border'. Thus the Shi-sumi are the four corner points of the Imperial palace (Fig. 3).[30] The Shi-sakai are the four borders of the Imperial Palace and also those of the Imperial City (Fig. 1). From these places, any pollution and any wicked spirits might possibly steal into the city; thus the Shinto priests held rituals and rites to prevent these things from penetrating the city. The Shi-sumi-Shi-sakai-no-Matsuri is held to expel and to purify the evil spirits at the four corners and the four borders of the palace and those of the City.[31]

At the Palace were held the rites of the Michiae-no-matsuri and the Chinka-sai.[32] The latter is the rite to prevent and to protect against fire. Both rites are held in June and December.[33] The picture (Fig. 4) is a scene from a tale (*Fudo-riyaku engi*) of exorcising the demons of disease. [34] It may be modelled on the ceremony of the Michiae-no-matsuri, in which the Shinto priests held a banquet to entertain the vessel-spectres with food and drink. The model for the figure on the right may have been Abe-no-Seimei, a famous Shinto priest and yin-yang diviner as well. The spectres are the allegorical figures that represent impurity and wicked spirits. They were believed to steal into the city hiding themselves in

---

[29] Koda (1981) 317-35; 'Shi-kaku-shi-kyo-no-matsuri' in the Kokushi dai-jiten VI (1985) 668.

[30] The rite held on 23 October, 914 is recorded by Minamoto-no Taka-akira (914-982). Seikyu-ki (1953) 28-9. Also the same rite held on 21 February, 1105 is recorded by Miyoshi Tameyasu in Choya-gunsai (composed in 1116). (1938) 378-9.

[31] The rites of the Shi-sumi and the Shi-sakai, held on the same day (8 June, 1351 are recorded in the Mibu-ke Monjo, IX (1987) 290-1.

[32] Ichijo Kanera (1402-1481) recorded the rites of Chinka-sai, Michiae-no-matsuri, Oharae in Kuji-kongen. Sekine (1903) 51-4.

[33] Koda, 318-21; Kokushi dai-jiten XIII. (1992) 350.

[34] Komatsu (1995) 96-97. Shinpo (1995) 137-47.

124

the vessels. The priests, by giving them a feast, prayed them not to enter into the city, and to withdraw from it.

Fig. 4

2) Religious Festivals, Ceremonies, and Rites

Furthermore, since about the beginning of the Heian Era, the priests came to think that the results of the curses of the dead should torment not only the nobles but also the townspeople. To them, epidemics and calamities derived from the curses of the dead, especially when they occurred just after the tragic, cruel, or unfortunate death of a famous person.[35] One of the reason, it is said, why Kanmu Tenno transferred the Imperial City from Nagaoka-kyo to Heian-kyo is because he was afraid of the curse of his brother, who died as he was suspected of plotting against Kamnu Tenno.[36] In order to calm the curses of the dead, the rites and religious observances were held lest the curses should do harm to the living.

Originally the rites for purification and those for appeasement were separate from one another. Gradually as time went by, the line between them grew dim. Thus they held religious observances to pray for the repose of their souls so that the dead might not bring a curse upon them, as well as so that the wicked spirits should not do harm to them.

---

[35] For example, Crown Prince Sawara-shinno (Sudo Tenno), Iyo Shinno, Fujiwara-no Kisshi, Tachibana-no-hayanari, Fun-ya-no Miyatamaro, Sugawara-no Michizane, and Minamto-no Taka-akira. Kawane (1991)233-4; Murayama (1967) 268-84; Hayami (1987) 94-6; Rekishi-chizu, 38-43. Fujiki (1959) 131-2.

[36] Rekishi-chizu, 16-21; Fujiki (1959) 40-2.

In the city of Heian-kyo various kinds of festivals, rites, and religious ceremonies and observances are held. Although town walls had not been constructed and the line drawn between the inside and outside of the city was unclear, the townspeople were conscious of the distinction: the inside of the city was purified and safe, while the outside was the opposite. Accordingly, around the city there were temples and shrines, and outside of the city there were funeral places and the barrier stations (Fig. 3). They were the border areas between the inside and the outside of the town, which were at the same time the border between the world of living people and that of the dead, this world of ours and the other world of the ghosts, wicked spirits, and the goblins. Thus, these places were dangerous, unsafe, and fearful.

The funeral places are found, for example, to the east (Toribeno at Higashi-yama); to the north (Rendaino); and to the west (Adashino). Additionally, there were other funeral places such as Yoshida-yama, Takeda, and Fukakusa (Fig. 3).[37] At the funeral places, those who died were left as they were, for example, at Adashino. Then, Kukai, a Buddhist priest, started Adashino Nenbutsu-ji, a temple, to pray to Amitabha for the dead.[38]

3) The Gion-matsuri Festival

In 863, an official ritual was held at the Shinsen-en Pond in the hope of calming the wicked spirits and removing the curse of the dead.[39] The ritual is called Goryo-e. On the other hand, privately, with the temples and the shrines as the core, religious rites and festivals, and memorial services were held. The Gion-matsuri was at first observed to appease and suppress a raging epidemic. Thereafter, it appears that a folk religion having to do with appeasing the avenging spirits of the departed came to be mingled with the rites of exorcism of the epidemic demons, as mentioned above. According to the tradition of the shrine, in 869, when the epidemic was raging, the Gion-sha shrine held the ritual, Gion Goryo-e, and the townspeople marched in procession with 66 *Hokos*, festival floats mounted with a decorative halberd, to Shinsen-en Pond. This is the beginning of the Gion-matsuri.

---

[37] Nishikawa, Kyoto. I. 34-6.

[38] Nishikawa, Kyoto. I. 36.

[39] Murayama, 268-78; Fujiki, 131-2, 281-88. The Shinsen-en Pond ('water of marvelous efficacy' in English) was a pond for the Tennos, following the concept of Chinese Imperial Gardens, where the nobles set boats afloat, played, and enjoyed composing poems and listening to music. The pond gradually became the place where the rites and the ceremonies prayed for the dead who died a tragic death, for example, by falling into a political snare. Nishikawa, Kyoto. I. 30-31.

126

§ 4. Literature and the Borders of the City

Because of the nature of the outside of the City, the border regions resulted in various kinds of legends and tales.

1) The Rokudo-Chinno-ji

Among such places there is a temple, the Rokudo-Chinno-ji (Fig. 3). *Rokudo* means the 'six roads' that lead to the six worlds: this world of ours and the other five worlds. According to religious teachings, all men are to go to the six worlds after death along the six roads. In Kyoto, near the Rokudo-Chinno-ji temple, it is said that there is a border that leads to the six roads. The temple has an interesting legend about Ono-no-Takamura, a man of letters in the ninth century. He is said to have worked at the office in the Palace Court in the daytime and still more to have worked every night at the office of Great Yama, the ruler of Hell as well as of the world of the dead.

2) Legends of Ogres

There are many legends and stories which are related to the ogres, who are said to live outside of the City. The Rajo-mon is especially notorious for ogres frequenting it. Once the gate was splendid and magnificent but it collapsed several times. After it fell down completely in 980, the gate was not reconstructed again.[40]

1. The Biwa of Murakami Tenno

A tale in the *Konjaku Monogatari-shu* (the Tales of Times Now Past), tells of an episode in the reign of Murakami Tenno (926-967; r.946-967) in the tenth century.[41]

One day, one of his *biwas*, a Japanese lute, was stolen. The same biwa, called Genjo, had a strange, supernatural power. When it was left dusty, it got angry and would not let itself be plucked. When a fire broke out at the Imperial Palace, the biwa was able to escape from the palace by itself without the help of men.

While thinking of his lost biwa, Minamoto-no-Hiromasa or Minamoto-no-Hakuga, a nephew of Murakami Tenno, happened to hear a melody played on the very same biwa. He was a famous court music player, an excellent biwa

---

[40] Nihon rekishi-chimei taikei (1979) 33-35.

[41] Konjaku monogatari-shu, XXIV-24. Konjaku monogatari-shu is the largest narrative stories in Japan compiled in the first half of the twelfth century. Sakakura (1978) I. 178-81; Komine (1994) IV. 430-2.

player, and is said to have taken lessons in the secret music of the biwa from Semimaru,[42] a legendary figure said to have been a famous biwa player as well as an excellent waka poet, one of whose waka poems is included in the *Hyakunin isshu*.[43]

Minamoto-no-Hiromasa walked along the Suzaku-oji, following the melody, and came to the Rajo-mon Gate. The biwa player was excellent and exquisite. He listened to playing and when music came to an end, he said, "Who are you who has played the biwa? The biwa you are playing is Mikado Murakami's. Today, hearing the sound of it, I have followed the sound and have come here." The player was an ogre and he returned the biwa to him. The biwa, bound with a cord, came down from above the gate. Receiving it, the nephew came back to the imperial palace with the biwa and returned it to Murakami Tenno.

3) The Legends of Modori-bashi

1. The Modori-bashi Bridge

In addition to the Rajo-mon, the Modori-bashi Bridge (Fig. 3) is also one of the famous places where one happens to encounter a supernatural wonder. The Modori-bashi lies at the northeastern corner of the Imperial Palace. The Modori-bashi was once a larger bridge over a wider and more massively flowing river than that of today. The bridge also lies at the border place. Near the bridge there is a shrine, Seimei Jinja, sacred to Abe-no-Seimei. He was said to have kept the twelve Shinsho, subordinates, who had supernatural powers, as his pawns. They had such terrible shapes and looks that his wife was afraid of them. This is why Seimei kept them at the foot of the bridge.[44]

2. Tale of Miyosi Kiyoyuki

Modori is a form of the verb *modoru*, 'to return', and *bashi* or *hashi* means 'bridge'. Thus the meaning of *Modori-bashi* is 'the bridge to and from which one is to go and return'. There is a story which tells how this bridge got its name.

Once, in the tenth century, a Buddhist priest, Jozo, who was leading an ascetic life at Kumano in Ki-i, the biggest peninsula in Japan, located on the southern part of the Kinki Region, was in a hurry on his way home in Kyoto because he had been told of his father's death. On the bridge, he met the funeral

---

[42] Konjaku monogatari-shu, XXIV-23.Komine IV (1994). 427-30; Sakakura (1978) I. 174-8.

[43] Hyakunin Isshu, No. 10. Ariyoshi, 52-5.

[44] Shinwa-densetsu (1963) 66; Shimazu (1929) 25.

procession of his father. He prayed and prayed very earnestly. Then, strangely enough, his father, Miyosi Kiyoyuki, a man of letters, came back to life. Since then the bridge was called the Modori-bashi.[45]

It is said that brides should not cross the bridge because brides who cross it would have their marriages annulled and would return to their parents' house. Whereas, on the other hand, soldiers during World War II, before going to the front line, willingly crossed the bridge so that they might return alive.[46]

3. Watanabe-no-Tsuna

The next story tells of the ogre who would sometimes appear near the bridge Modori-bashi.[47] Once, in the tenth century, there lived Watanabe-no-Tsuna, a Samurai or Bushi, a Japanese warrior. He was one of the four best retainers of Minamoto-no-Yorimitsu (or Minamoto-no-Raiko). One night when he was riding near the Modori-bashi, at the foot of the bridge, a beautiful young lady was standing alone. She asked Tsuna to escort her to her house near the Gojo (Fifth Street) in the city. Then Tsuna had her mount his horse and he walked in the direction indicated. On the way she told him that, because her house was outside the city she would like Tsuna to escort her to her house. Tsuna agreed and began to go ahead.

In truth, the lady was an ogre. It clutched hold of Tsuna's *motodori*, his topknot, on the way and tried to fly away to Mt. Atago-yama (Fig. 1), which was the den of the ogre. That night Tsuna had his lord Raiko's famous sword, Higekiri, with him. At once he cut off the arm of the ogre with the sword. The ogre ran away without the arm.

Tsuna, obeying the advice of Abe-no-Seimei, shut up the arm in a box by incantation and he himself observed abstinence and stayed home for seven days. On the sixth day, a woman who was announced as Tsuna's foster mother visited him. He received her against his will. The woman then asked him to show her the arm of ogre. Also against his will he opened the box and showed it to her. The woman was, in reality, the ogre. She gazed and gazed at the arm. And then, terrible to say, suddenly she changed back into her original shape as the ogre. No sooner had the monster snatched the arm than it ran away, breaking through the gable. Since then they say that the Watanabe family has never built houses with gables.

---

[45] Tokoro (1989) 204-6.
[46] Moriya, 20-21.
[47] Mizuhara (1988) I. 62-5; Shimazu (1929) 21-25.

3. Ibaraki-doji and Shuten-doji

In another version, for example, in Yo-kyoku (Noh song) and other versions, Watanabe-no-Tsuna cut off the ogre's arm in front of the Rajo-mon.[48] In this version the story is as follows: one day, soon after having destroyed Shuten-doji, a notorious bandit, and his followers, Raiko and his four big vassals, along with a guest, Hosho, had a drinking party. During the party they happened to talk about the ogre that had recently often appeared around the Rajo-mon. Then Watanabe-no-Tsuna decided to make certain of the rumour and went out. When he entered the gate, he encountered the ogre. After fighting against the ogre he cut off its arm. The ogre is Ibaraki-doji and later he also changed his appearance to come to Tsuna's house to snatch back his arm.

In another version, the young lady who stood at the foot of the Modori-bashi is Hashi-hime of Uji (Princess of the Bridge of the Uji) (Fig. 1).[49] According to this version, the story involves a couple. As the husband took a second wife, the first wife became mad with jealousy. Then at midnight she prayed a special prayer for a curse in the hope of casting the curse on the others. Thus she became an ogre and did harm to the townspeople.

§ 5. The Civil War and the Modernised Kyoto

1) The Civil War

The Heian Era came to an end with the fall of the Heike Clan, and Minamoto-no-Yoritomo established the seat of the shogunate in Kamakura in 1192 (-1333). The government (the Kamakura Bakufu) was at Kamakura (Fig. 2) but the Imperial City was still Kyoto. After the Kamakura Shogunate, the Ashikaga Clan established the seat of the shogunate in Muromachi in Kyoto, and the government of that period (the Muromachi Bakufu) is called the Muromachi Shogunate (1336-1573).

At the end of the Muromachi Era, the Imperial City of Kyoto fell upon hard times: the Onin-Bunmei-no-ran (1467-77), the Civil War, broke out; Kyoto was destroyed, burnt, and laid waste. One third of Kyoto was burnt down and the town became disordered, unsafe, and dangerous. The lamentable sight was depicted in a waka poem.

---

[48] Nogami (1985) V. 529-38.
[49] Shimazu (1929) 25-30.

130

Nareya shiru
Miyako ha nobeno Agehibari
Noboru wo mitemo
Otsuru namidaha[50]

Can you imagine?
How our Miyako is devastated and ruined;
Once such a beautiful Castle City.
Seeing a skylark soaring up in the sky,
Tears are falling in drops.

During the war, many of the nobles and high-ranked priests left Kyoto and stayed in the provinces, where, as a result, culture in the manner of Kyoto was nurtured. For example, various kinds of regular annual events in the Imperial Court became popular, and the common people also imitated the nobles in performing them.[51]

After the Onin-Bunmei-no-ran, the nobles lost the power and energy to make the burnt-out Kyoto arise from its ashes, but the townspeople managed to do it. The Gion-matsuri was revived in 1500, which was the symbol of their joy and mettle.[52] Their power was, however, checked by the ambition of the feudal lords of the Warring States Period (1467-1568) in the provinces of Japan, who were the rising class of the period, coming up to Kyoto to rule over the whole country.

Among them, Oda Nobunaga came near to the supreme position. He gained mastery over any movements against him. But his ambition failed when the Raid on the Honno-ji Temple took place in 1582, in which Oda Nobunaga was killed by Akechi Mitsuhide, his retainer. Toyotomi Hideyoshi, also one of Nobunaga's retainers, destroyed Mitsuhide and was unrivalled among his peers. He embarked upon the reconstruction of the city of Kyoto with a great reorganisation.

2) The 'Rakuchu-rakugai-zu' and the 'Nanban-ji'

From the end of the Muromachi Era to the beginning of the Edo Era, the depictions of the new-born city began to change greatly. Views of the city appeared in paintings on folding screens, *byobu*, called "Rakuchu-rakugai-zu." From near the end of the Warring States Period, Rakuchu-rakugai-zu were painted quite frequently.[53] The feudal lords vied against one another to buy the

---

[50] Onin-ki. 409-412, esp. 412.

[51] Nishikawa, Kyoto. I. 86-7.

[52] Nishikawa, Kyoto. I. 88-9.

[53] Moriya, 96-100, 104-8, 110, 113, 120-1, 129. For editions of the 'Rakuchu-rakugai-zu', Cf. the Machida-ke (1987) Version, the Funaki-ke (1987) Version, and the Uesugi-ke (1987) Version.

byobus entitled "Rakuchu-rakugaizu." *Raku-chu* refers to the inside of Kyoto, and *raku-gai* the outside of it. Among the motifs of the paintings, one of the new images in Kyoto is the *Nanban-ji*.

In 1543, a ship from Portugal was washed ashore on a small island in the south of Japan, Tane-ga-shima. These were the first Portuguese to visit Japan. In 1549, Saint Francisco Javier (Francis Xavier) visited Japan. Thereafter, Xavier led his missionary work from Kyushu, the southern part of Japan and in 1550 he came to Kyoto. In 1551, the first Christian church was built in Yamaguchi in the western part of Japan. In 1569, a Christian church was built in Kyoto with the permission of Oda Nobunaga. In 1576 the church was reconstructed into a three-storied one to conserve space. In 1587, in accord with the ban on Christianity by Toyotomi Hideyoshi, the church in Kyoto, as well as other 53 churches, was destroyed.[54]

This church was called *Nanban-ji* in Japan. *Ji* means 'temple'. To the Japanese in those days, a church was a kind of a temple. A Nanban-ji is painted in a Rakuchu-rakugai-meisho-senmen-zu (a fan-shaped picture), a variation of the Rakuchu-rakugai-zu, in which the characters *nanbaudau no zu* can be found, that is, 'the picture of Nanban-ji'. In those days the building was so new, so unfamiliar, and so uncommon that a great many onlookers came to see it.

3) Odoi

Among Hideyoshi's plans, a notable one is the *Odoi* (Fig. 3), a kind of dike, which suggests that the new ages are coming near. The dike of about 22.5 kilometers in length was constructed in 1591 and enclosed the City.[55] On the top of the Odoi, Hideyoshi's had a bamboo thicket planted. Though Heian-kyo was constructed and planned after the model of a Chinese castle town, the town wall had not been built except for the gate Rajo-mon. It was Hideyoshi who had the wall constructed.

Thus, in Japan, the town wall was constructed in the early modern period, whereas in Europe town walls were built in ancient times or in the medieval period. When the Odoi was to be constructed, some nobles reportedly said that the dike would be unnecessary for their city of Miyako.[56] In fact, it seems that the townspeople of Heian-kyo did not become accustomed to the Odoi. Gradually there appeared tears and rents here and there in it, and some parts of it were destroyed.

---

[54] Nishikawa, Kyoto. I. 96-7.
[55] Nishikawa, Kyoto. II. 14-7. Ashikaga, 68-9; Moriya, 108-9.
[56] Nishikawa, Kyoto. II. 16.

Closing

After Hideyoshi, Tokugawa Ieyasu assumed power; he established the seat of the shogunate in Edo (Fig. 2) in 1603 (-1868), today's Tokyo, that is, the Edo Bakufu or the Tokugawa Bakufu, although the Imperial City was still Kyoto.

There is one thing to be noted about Edo. The city is said also to be constructed according to the idea of Heian-kyo, the idea of the Four-Gods principle.[57] Thus, on the east, there stands Seiryu, the Hirakawa River; on the south, Suzaku, Edo-minato, the harbor; on the west, Byakko, a trunk road, Tokai-do; and on the north, Mt. Fuji.

Today Tokyo is generally regarded as the capital of Japan. There is a view, however, that the capital of Japan was, has been, and is still Kyoto. The first reason is that the Takamikura, the Imperial Throne, is even now in the Imperial Palace of Kyoto. Another reason is that the imperial edict of the transfer of the capital to Tokyo has not been officially proclaimed.[58] Apart from the argument, for most Japanese, Kyoto is a special city, the eternal city of Japan, as Rome is the eternal city of the world.

---

[57] Naito-Akira (1982) I. 12-15; Naito-Masatoshi (1987) 79-81.
[58] Kokushi dai-jiten. X. 42.

Bibliography

I. Sources

Akimoto, Kichiro, ed. *Fudoki* (*The Provincial Historical Records*). *Nihon koten-bungaku taikei 2* (*An Anthology of Classical Japanese Literature 2*). Tokyo: Iwanami shoten, 1958; 1984.

Ariyoshi, Tamotsu. *Hyakunin isshu*. Tokyo: Kodansha, 1983.

Asao, Naohiro, et. al. eds. *Kyoyo-fu no rekishi* (*A History of Kyoto-Prefecture*). Tokyo: Yamakawa Shuppan-sha, 1999.

Ashikaga, Kenryo, ed. *Kyoto rekishi atorasu* (*Historical Atlas Kyoto*). Tokyo: Chuo-koron-sha, 1994.

Chamberlain, Basil Hall, tr. *The Kojiki: Records of Ancient Matters*. First printing, Yokohama: Lane, Crawford, 1883; Rutland, Vermont & Tokyo, Japan: Charles E. Tuttle Company, 1981.

Endo, Teruaki, ed. *Toshi wo kangaeru* (*Thinking of the Cities*). Yokohama: Yurindo, 1985.

Fujiki, Kunihiko. *Nihon zenshi III: Kodai 2* (*A Complete History of Japan III: Ancient Times 2*) Tokyo: Tokyo-Daigaku shuppan-kai, 1959.

Fujiwara, Yusetsu, ed. *Jogu Shotoku-taishi den hoketsu-ki* (*A supplement of Jogu Shotoku-taishi*) *Shotoku-taishi Zenshu II* (*The Complete Words of Shotoku-taishi II*) General editor, Shotoku-taishi hosan-sai. Kyoto: Rinsen shoten, 1944; 1988.

Groot, J. J. M. de. *The Religious System of China*. 6 vols. Leyden: Brill, 1892-1910.

Hayami, Tasuku. *Jujutsu shukyo no sekai* (*The World of Religious Incantation*). Tokyo: Hanawa shobo, 1987.

Hayashiya, Tatsusaburo. *Rekishi no naka no toshi* (*The Cities in History*). Tokyo: Nippon-hoso shuppan kyokai, 1973; 1982.

Ichijo Kanera. *Kuji-kongen* (*A Commentary of Annual Official Events, and Military Practices and Usages of the Imperial Palace Court*). Ed. Masanao Sekine. Tokyo: Rokugo-kan, 1903; Dai-ichi shoten, 1986□Original manuscript is about 1423.

Ishida, Mizumaro. *Reibun Bukkyo-go dai-jiten.* (*A Comprehensive Dictionary of the Terms of Buddhism with Examples*). Tokyo: Shogaku-kan, 1997.

Kanaoka, Shuyu, ed. *Kukai Jiten* (*A Dictionary of Kukai*). Tokyo: Tokyo-do, 1979.

Kawane, Yoshiyasu 'Go-ryo On-ryo toha nanika' ('What are the Avenging Spirits of the Departed and the Vengeful Spirits?'). *Soten, Nihon no rekishi: Dai III kan Kodai-hen 2* (*Points in Dispute in Japanese History III: Ancient Times 2*). Ed. Takehiko Yoshimura & Masayuki Yoshioka. 6 vols. Tokyo: Shin-jinbutsu orai-sha, 1991. 226-35.

Kishi, Toshio. *Nihon no kodai-kyuto* (*Capitals of Ancient Japan*). Tokyo: Nippon-hoso shuppan kyokai, 1981.

Koda, Toshio. 'Shikaku-sai ko' ('A Study of Shikaku-sai'). *Heian-cho rinji kuji ryakkai. (A Commentary of the Extra Official Government Affairs and the Ceremonies in the Heian Era)*. Tokyo: Zoku-Gunsho-ruiju kansei-kai, 1981. 317-35.

Kodama, Kota, ed. *Shiryo ni yoru Nihon no ayumi: Kodai-hen* (*A Japanese History from the Historical documents*). Tokyo: Yoshikawa kobun-kan, 1960.

Kojima, Noriyuki & Eizo Arai, eds. *Kokin waka-shu.* (*An Anthology of the Kokin waka-shu*). *Shin-Nihon koten bungaku taikei 5* (*A New Anthology of Classical Japanese Literature no. 5*). Tokyo: Iwanami shoten, 1989.

Komatsu, Shigemi, ed. *Zoku-zoku Nihon emaki taisei: denki, engi hen 4. Hoyake Amida engi, Fudo-riyaku engi. (Third Series of an Anthology of Japanese Picture Scrolls: Biographical Stories and the Foundations of the Temples and the Shrines 4)*). Tokyo: Chuo koron-sha, 1995.

Komine, Kazuaki, ed. *Konjaku monogatari-shu IV* (*Tales of Times Now Past*) *Shin Nihon koten bungaku taikei 36* (*An Anthology of Classical Japanese Literature no. 36*). Tokyo: Iwanami shoten, 1994.

Kubota, Jun et. al. eds. *Gaisetsu, Nihon bungaku-shi* (*General Sketch: Hystory of Japanese Literature*). Tokyo: Yuhikaku, 1979.

Kurano, Kenji & Yukichi Takeda, eds. *Kojiki, Norito* (*The Kojiki and the Shito Prayers*). *Nihon koten bungaku taikei 1* (*An Anthology of Classical Japanese Literature no. 1*). Tokyo: Iwanami shoten, 1958; 1966.

Mesaki, Shigekazu. *Zusetsu Fusui-gaku* (*Illustrated Lectures on the Feng Shui*). Tokyo: Tokyo shoseki, 1998.

Minamoto-no Taka-akira. *Seikyu-ki II* (*The Records by a Noble Lived in the West Palace II*). Shintei zoho kojitsu-sosho (*A New and Expanded Edition of an Anthology of the Studies in Ancient Court, and Military Practices and*

Usages). Original manuscript is in the tenth century. Ed. Kojitsu sosho henshu-bu. Tokyo: Meiji-tosho & Yoshikawa kobun-kan, 1953.

Miyoshi, Tameyasu, ed. *Choya-gunsai* (*A Miscellany*). *Shintei Zoho Kokushi taikei, dai 29 kan, jo* (*New Edition of an Anthology of the Historical Documents of Japan: Vol. 29, Part One*). Original manuscript is in 1116. Ed. Katsumi Kuroita. Tokyo: Yoshikawa-kobun-kan, 1938; 1964; 1999.

Mizuhara, Hajime, ed. *Shintei Genpei josui-ki Dai 1 kan* (*New Edition of the Taira-Minamoto War*). Tokyo: Shin-jinbutsu orai-sha, 1988.

Moriya, Katsuhisa & Mitsuo Inoue. eds. *Heian-kyo 1200-nen.* (*Heian-kyo 1200 Years*) Kyoto: Tanko-sha, 1994.

Murayama, Shuichi. 'Heian bukkyo no tenkai' ('The Development of Buddhism in the Heian Era') *Nihon bukkyo-shi: I Kodai-hen* (*The History of Buddhism in Japan: 1. Ancient times*). Ed. Saburo Ienaga. Kyoto: Hozo-kan, 1967. 241-98.

Naito, Akira. *Edo no machi, jo: Kyodai-toshi no tanjo* (*Towns of Edo, part one: the Birth of a Megalopoli*s). Tokyo: Soshi-sha, 1982.

Naito, Masatoshi. 'Toshi no sei-kukan to takai: Edo no Ju-teki kosumoroji' ('The Sacred Space and the Other World: The Incantatory Cosmology of Edo') *Bukkyo minzoku-gaku taikei 3: Seichi to takai-kan* (*An Anthology of Buddhism Folklore 3: Sanctuary and Thought of the Other World*). Tokyo: Meicho-shuppan, 1987. 79-106.

Nakamura, Hajime. *Bukkyo-go dai-jiten* (*A Dictionary of the Terms of Buddhism*). A compact edition. Tokyo: Tokyo shoseki, 1981; 1983□

______. *Zusetsu Bukkyo-go dai-jiten* (*An Illustrated Comprehensive Dictionary of the Terms of Buddhism*). Tokyo: Tokyo shoseki, 1988.

Naramoto, Tatsuya, et. al. *Kyoto hyaku-wa: Kyoto sen-nen heno Annai* (*One Hundred Episodes of Kyoto: A Guide to Kyoto of One Thousand Years*). Tokyo: Kadokawa shoten 1984.

Nishikawa, Koji. *Toshi no shiso* (*A Thought on the Cities*). Tokyo: Nippon-hoso shuppan kyokai, 1982.

Nishikawa, Koji and Toru Takahashi. *Kyoto 1200 nen* (*Kyoto 1200 Years*). 2 vols. Tokyo: Soshi-sha, 1997.

Nogami, Toyoichiro. *Yo-kyoku Zenshu: Maki 5, Rasho-mon* (*A Complete Collection of the Noh Songs: Book Five, Rasho-mon*). Tokyo: chuo-koron-sha, 1985.

Porter, William N., tr. *A Hundred Verses from Old Japan*. Rutland: Charles E. Tuttle, 1909; 1994.

Sakakura, Atsuyoshi, Giken Honda & Yoshiaki Kawabata, eds. *Konjaku monogatari-shu 1*. (*Tales of Times Now Past, Bk. I*) *Shincho Nihon koten shusei (Shincho Anthology of Classical Japanese Literature*). Tokyo: Shincho-sha, 1978; 1986.

Sakamoto, Taro, Saburo Ienaga, Mitsusada Ino-ue & Susumu Ono, eds. *Nihon-shoki*, jo (*Nihon-shoki, Part 1*). *Nihon koten-bungaku taikei 67 (An Anthology of Classical Japanese Literature 67*). Tokyo: Iwanami shoten, 1967; 1978.

Satake, Akihiro, Hideo Yamada, Rikio Kudo, Masao Otani & Yoshiyuki Yamazaki, eds. *Man-yo-shu 1* (*Man-yo-shu 1*) *Shin-Nihon koten bungaku taikei 1* (*A New Edition of Classical Japanese Literature 1*). Tokyo:Iwanami shoten, 1999.

Shimazu, Hisamoto. *Rasho-mon no oni* (*Ogre of the Rasho-mon*). Tokyo: Shincho-sha, 1929; Tokyo: Heibon-sha, 1975.

Shinpo, Toru. '"Fudo-riyaku engi"ni tsuite' ('On "Fudo-riyaku engi"'). *Hoyake Amida engi, Fudo-riyaku engi. Zoku-zoku Nihon emaki taisei: denki, engi hen 4*. Ed. Shigemi Komatsu. Tokyo: Chuo koron-sha, 1995. 137-47.

Takahashi, Shunjo. 'Shugei-shuchi-in ni tsuite' ('On Shugei-shuchi-in') *Ronshu, Kukai to Shugei-shuchi-in: Kobo-daishi no kyoiku, jo* (*A collection of essays on Kukai and Shugei-shuchi-in: the Education of Great Teacher Kobo-daishi, I*). Ed. Yukio Hisaki & Kazuo Oyamada. Kyoto: Shibun-kaku, 1984. 43-54.

Takikawa, Masajiro. *Hosei-shi ronso Dai 2 satsu: Kyo-sei narabini tojo-sei no kenkyu* (*A Collection of the History of Law II: A Study of the Law System of the Imperial City*). Tokyo: Kadokawa shoten, 1967.

Takinami, Sadako. *Heian kento* (*The Construction of Heian-kyo*). Tokyo: Shuei-sha, 1991.

Tokoro, Isao. *Miyoshi Kiyoyuki*. Tokyo: Yoshikawa kobun-kan, 1989.

Yajima, Genryo. 'Shugei-shuchi-in wo megurite' ('Concerning Shugei-shuchi-in') *Ronshu, Kukai to Shugei-shuchi-in: Kobo-daishi no kyoiku, jo*. Ed. Yukio Hisaki & Kazuo Oyamada. Kyoto: Shibun-kaku, 1984. 245-59.

II. Secondary Literature

Ajia rekisi-jiten (*A Dictionary of Asian History*). Ed. Kunihiko Shimonaka. Tokyo: Heibon-sha, 1962; 1984.

Kokushi dai-jiten (*A Comprehensive Dictionary of Japanese History*). Ed. by the Compilation Committee of *a Comprehensive Dictionary of Japanese History*. Tokyo: Yoshikawa kobun-kan, Vol. VI. 1985; Vol. X. 1989; Vol. XIII. 1992; Vol. XIV. 1993.

Kyoto no rekishi 3: Kinsei no taido (*The History of Kyoto 3: the Stirrings of the Early-Modern Times*). By Kyoto-shi. Kyoto: Gakugei shorin, 1968.

Kyoto no rekishi to bunka: Kyoto bunka hakubutsu-kan rekishi tenji annai. (*History and Culture of Kyoto: a Guide to the Historical Exhibition of Kyoto Cultural Museum*). Ed. Kyoto bunka hakubutsu-kan gakugei dai 2 ka. Kyoto: Kyoto-fu Kyoto bunka hakubutsu-kan, 1988; 2004.

Mibu-ke Monjo IX (*The Records Inherited in the Mibu Family No. 9*). *Zusho-ryo sokan* (*The Imperial Library Series*). Ed. the Imperial Library. Tokyo: Meiji shoin, 1987.

Nihon kodai-shi dai-jiten (*A Comprehensive Dictionary of Ancient Japan*). Ed. Masaaki Ueda. Tokyo: Daiwa shobo, 2006.

Nihon rekishi-chimei taikei, dai 27 kan: Kyoto-shi no Chimei (*An Anthology of Historical Place Names of Japan Vol. 27; The Place Names of Kyoto-shi*). Ed. Kunihiko Shimonaka. Tokyo: Heibon-sha, 1979.

Onin-ki. Gunsho ruiju, dai 20 shu, Kassen (*A Collection of Various Books According to Category, Vol. 20, Battles*). Ed. Hanawa Hoki-ichi. Tokyo: Zoku Gunshoruiju kanse-kai, 1929: Third edisiton, 1959. (no new line after "Zoku")

Rakuchu-rakugai-zu taikan (*A Broad Perspective Edition of a Folding Screen of the Grand View of Kyoto*). *Uesugi-ke bon.* Ed. Hisatoyo Ishida, Akira Naito & Katsuhisa Moriya. Tokyo: Shogaku-kan, 1987.

Rakuchu-rakugai-zu taikan, Machida-ke bon. Ed. Hisatoyo Ishida, Akira Naito & Katsuhisa Moriya. Tokyo: Shogaku-kan, 1987.

Rakuchu-rakugai-zu taikan, Funaki-ke bon. Ed. Hisatoyo Ishida, Akira Naito & Katsuhisa Moriya. Tokyo: Shogaku-kan, 1987.

Rekishi-chizu-hon: Shitte tazuneru Kyoto (*A Book of Historical Map: A guide to Studying Kyoto*). Ed. Rekishi tanbo kenkyu-kai. Tokyo: Daiwa shobo, 2006.

Sekai-isan to toshi (*World Heritages and Cities*). Ed. Nara Daigaku bungaku-bu Sekai-isan kosu. Nagoya: Fubai-sha, 2001.

Shinwa Densetsu jiten (*A Complete Dictionary of Myth and Legends*). Ed. Haruhiko Asakura, Shoji Inoguchi, Hirohiko Okano & Ken Matsumae. Tokyo: Tokyo-do, 1963; 1997.

Appendix

Chronological Table from the Heian Era to the Beginning of the Early Modern Times

784	Enryaku 3	Kanmu Tenno transferred the capital to Nagaoka.
794	Enryaku 13	Kanmu Tenno transferred the capital to Kyoto, Heian-kyo.
816	Konin 7	The Rajo-mon fell for the first time.
863	Jogan 5	Goryo-e is held at the Shinsen-en Pond.
869	Jogan 11	Rising of an epidemic. The first festival of Gion-matsuri was held.
960	Tentoku 4	Dairi, the Imperial Palace, was burnt down for the first time (in the reign of Murakami Tenno).
976	Jogen 1	Dairi, the Imperial Palace, was burnt down.
980	Tengen 3	The Rajo-mon collapsed. It has not been reconstructed since then.
1005	Kanko 2	Dairi, the Imperial Palace, was burnt down.
c.1008	Kanko 5	At least a part of the Tale of Genji was completed when Fujiwara-no Michinaga was Regent.
1052	Eisho 7	The first year of Mappo thought (the latter days of Buddhism). Fujiwara-no Yorimichi changed his second mansion at Uji into a temple and called it Byodo-in.
1053	Eisho 8	Ho-o-do (a temple) of Byodo-in was completed.
1156	Hougen 1	The Hogen Civil War, *Hogen no ran*, broke out.
1159	Heiji 1	The Heiji Civil War, *Heiji no ran*, broke out.
1167	Nin-an 2	Taira-no Kiyomori became the Grand Minister.
1177	Jisho 1	A big fire, Taro-Fire, broke out in Kyoto. The Main Hall of the Imperial Palace was burnt down. Since then it has not been reconstructed.
1178	Jisho 2	A big fire, Jiro-Fire, broke out in Kyoto.
1180	Jisho 4	A big fire broke out in Kyoto. Taira-no Kiyomori transferred the capital to Fukuhara, today's Kobe.
1185	Bunji 1	Battles of Yashima and of Danno-ura, The fall of the Heike Clan. Antoku Tenno drowned.

1192	Kenkyu 3	Minamoto-no Yoritomo established the seat of the shogunate in Kamakura (Kamakura Era). Still, the capital city was Kyoto.
1227	Antei 1	Dai-dairi, the Imperial Palace, was burnt down because of a big fire. Since then it has not been reconstructed.
c.1235	Katei 1	Hyakunin isshu was compiled by Fujiwara-no Teika.
1274	Bun-ei 11	The first Mongolian Invasion.
1281	Ko-an 4	The second Mongolian Invasion.
1334	Kenmu 1	The lampoon of Nijo-gawara was written.
1338	Engen 3 · Ryakuo 1[59]	Ashikaga Takauji became Seii-taisho-gun, the 'barbarian-subduing great general'.
1378	Tenju 4 · Eiwa 4	Shogun Yoshimitsu moved to the new Palace at Muromachi, called the Flowery Palace.
1397	O-ei 4	A Ceremony for the construction of Kinkaku-ji (a palace of Shogun Yoshimitsu) was held.
1467	Onin 4	The Onin War broke out (- 1477). Many nobles went out to the provinces.
1483	Bunmei 15	Shogun Yoshimasa moved to Ginkaku-ji (a palace of Shogun Yoshimasa).
1543	Tenbun 12	A ship from Portugal washed ashore at Tane-ga-shima.
1569	Eiroku 12	Nobunaga permitted Luis Frois to live in Kyoto.[60]

Prof. Yuko Tagaya
Kanto Gakuin University
3-22-1 Kamariya-minami, Kanazawa-ku,
236-8502 Yokohama, Japan
E-Mail: tagaya@kanto-gakuin.ac.jp

---

[59] During 1336 and 1392, The Imperial Court was divided into two: the Northern and Southern Dynasties, each of which had its own Imperial era name.

[60] Luis Frois, Portuguese, is a Jesuit priest who came to Japan in 1563 and died at Nagasaki in 1597.

A Medieval Poet's Sense of Humour: Oswald von Wolkenstein and Emperor Sigismund in Paris

While Rome only recovered from its medieval decline at the end of the fifteenth century, Paris had grown steadily from Merovingian times onward. Paris's location at a crossroads between land and river trade routes in areas of abundant agriculture had made it one of France's principal cities by the tenth century, rich in royal palaces, wealthy abbeys, and a cathedral. By the twelfth century, Paris had become one of Europe's foremost centres of learning and the arts. Despite the Hundred Years War between the English and French kings, the growth of the capital accelerated considerably during the fourteenth century because of the combined efforts of King Charles V (1338-80), his brothers, the 'Princes des fleurs de lis', and his son, King Charles VI.

Paris in the Reign of King Charles VI

As Bernard Guenée, one of the foremost specialists of this period, put it (in: *Paris 1400*, p. 19): 'After the restorative reign of Charles V, who died in 1380, France experienced for more than 20 years under Charles VI a period of peace and exceptional prosperity (...) from which Paris received the greatest benefit.'

In order to remind the great public of this forgotten time of glory in the history of the French capital, in 2004 a magnificent exhibition was organised in the Louvre with the title "Paris 1400 – Les arts sous Charles VI". Although most of the art treasures of this period had vanished, the remaining examples bore convincing witness to the extraordinary wealth and growth of the French capital at the turn of the fourteenth to the fifteenth century. In addition, the presentation of recent archaeological research, excavations, and restorations aided our understanding of how Paris had grown to become the largest city in Europe, expanding its surface area to more than 400 hectares (990 acres) and its population to some 200 000 inhabitants.

But not only sheer quantity had promoted Paris to the leading position amongst European capitals: there was also the exceptional concentration of high-ranking elites that had gathered in the French capital at that time. The great number – more than a dozen royal households – alone attracted artists, mainly from Northern Europe. Moreover, the German Queen-Consort, Isabeau of Bavaria, added a considerable number of German noblemen to her entourage including her brother Duke Eberhard of Bavaria. In addition to the international composition of the Parisian aristocracy, the financial sector equally attained international dimensions, as it was largely dominated by bankers from Italy.

These sections of the population, the royalty and nobility, merchants and tradesmen, artists and craftsmen, including the officers of the guilds and the city government, occupied most of the quarters on the right bank of the river Seine as well as the 'City Island', the oldest part of Paris. The left bank, called the Latin Quarter, in contrast, was characterised by a domination of clergymen and all sorts of learned people who had been attracted from all over Europe to gain their degrees or to teach at one of the renowned Parisian Schools. By the end of the fourteenth century, thanks to its glorious past and the presidency of its highly renowned Chancellor Jean Gerson (1363-1429), one of those schools, the Sorbonne, had become the largest and most prestigious university in Europe.

Emperor Sigismund of Luxembourg and his Visit to Paris in 1416

In 1416, the year when Sigismund of Luxembourg paid King Charles VI a visit, history granted yet another superlative to the French capital.

Born in 1368, the third son of Holy Roman Emperor Charles IV, Sigismund had turned out to be the most fortunate of the emperor's numerous children. After his early marriage to the Hungarian crown princess Mary, Sigismund became King of Hungary. Despite the death of his consort and several upheavals against his royal regime, Sigismund managed to maintain himself in power for 50 years, a record in Hungarian history.

Just as he had done in his Hungarian Kingdom, Sigismund emerged as one of the most skilful rulers of the Holy Roman Empire. Thus, in 1411, Sigismund succeeded in acquiring the necessary votes of the German Electors to succeed Emperor Rupert of Germany. Besides his superior talents as a diplomat and politician, Sigismund also demonstrated an unusual amount of energy, persuasiveness, and strategic planning when he took over the task of ending the Great Western Schism that had resulted from the Avignon Papacy and divided Western Christendom since 1378.

To start with, Sigismund obtained a promise from Antipope John XXIII in Pisa that a council should be called at Constance in 1414 to settle the Western Schism. Right from the beginning, Sigismund fully assumed his new position as *advocatus et defensor ecclesiae* in making sure that delegations from all parts of the Latin and even orthodox Church gathered in the German city at Lake Constance, a place that he had chosen for its symbolic name (W. Brandmüller, 1999, 129-133). The gathering turned out to be the largest assembly in medieval Church history and promoted Sigismund to a position of supreme power: the highest-ranking Western sovereign and spiritual leader of the Latin Church.

Consequently, Sigismund took a leading part in the deliberations of this assembly. But when he realised that he was not able to persuade the delegates of the Avignon obedience to agree to the abdication of their Pope, he decided to make a journey to Spain in order to discuss the matters personally with Antipope Benedict and the leading Spanish King, Ferdinand I of Aragon (W. Brandmüller, 1999, 401-408). So in 1415, Sigismund quit the sessions in Constance and set out for Perpignan with an entourage of more than 300 people (Hartmann, 1997, 137). After several months of extremely difficult negotiations, first in Perpignan, which belonged to the Kingdom of Aragon, then in Narbonne and Avignon, the King of the Romans obtained a partial success sealed in the so-called agreement of Narbonne (*Capitula Narbonensia*), by which all Spanish kings agreed to withdraw from Pope Benedict's obedience (W. Brandmüller, 1997, 39-54).

While the good news of Sigismund's diplomatic success spread quickly, cities such as Avignon and Lyon celebrated the King of the Romans and his entourage with triumphal receptions, splendid banquets, balls and tournaments. At one of these occasions, probably in Lyon, messengers of the French King Charles VI called on the designated Emperor. The envoys mainly asked Sigismund whether he would agree to meet his 'cousin' in Paris in order to negotiate a peace treaty between Charles VI and King Henry V of England (1387-1422) (J. K. Hoensch, 1997, 225). At least, that is the way German historians such as Sigismund's most recent biographer, J. K. Hoensch, try to explain how and why the German sovereign paid a visit to King Charles VI in Paris.

Most present-day French historians and the French chroniclers of the fifteenth century, however, give a different explanation. And if we turn our eyes to a German witness and member of Sigismund's entourage such as Oswald von Wolkenstein, the whole story is told in yet another version, concealed in highly subjective, intriguingly humorous poetry. In addition, in each case, the perception of Paris, its importance and characteristics are reflected in differing images, thus giving us authentic examples of medieval perceptions of the other and of the self.

The Royal Chronicler of Saint-Denis and his Report about the Emperor's Visit

First we examine the Latin account of Sigismund's visit to Paris taken from the official chronicle of the French Kingdom. In the past, its author simply had been called the "Religieux de Saint-Denis". Recently he has been identified as Michel Pintoin. At any rate, the chroniclers of Saint-Denis are all held to be well informed and to bear authentic witness to the events on which they were reporting.

In the book about the reign of Charles VI, the chronicler devotes chapter 39[1] to the sumptuous reception offered by the city of Paris to the King of the Romans. He first gives the following explanation as to why the German ruler received his invitation:

> Meanwhile his Majesty the King of the Romans left Narbonne and set out for France. For a long time already, he had had a deep desire to meet the King of France and the Duke of Berry, his most beloved cousins, and to spend some days in their pleasant company. (Edition L. Bellaguet, 1844, 743 – my translation).

This opening sentence is striking, not only because the royal chronicler tells us that it was the King of the Romans who wished to meet the French king but because the whole trip is characterised as a private visit, as if Sigismund wanted to take some days off to see his relatives at leisure.

In fact, the whole report does not tell us much about the political purpose of Sigismund's visit. Instead the chronicler gives us a description of the visitors' programme and the high honours the designated Emperor and his entourage had been offered at the expense of the French crown.

The programme comprised the following points:

March first 1416 – splendid escort composed of all high-ranking officers into the city and the Louvre, the accommodations for the imperial entourage. Because of the illness of the French King, the King of the Romans used his leisure time to visit the following places:

1) the 'glorious University of Paris', where he observed the official acts of its faculties;
2) the House of Parliament, where he attended hearings;
3) the royal residences outside Paris;
4) the Abbey of Saint-Denis, where he stayed for 10 days.

In addition to the sight-seeing programme, several sumptuous banquets had been organised, mostly by Queen Isabeau and the Duke of Berry to honour their

---

[1] L. Bellaguet (ed. and trans.): *Chroniques du Religieux de Saint-Denys contenant le Règne de Charles VI, de 1380 à 1422*, Paris 1844, Livre XXXVIe, Kap. 39, vol. 5, p. 743-747. The anonymous clerical editor of the Latin series has recently been identified as Michel Pintoin; cf. P. Bourgain: Chronik. E. Frankreich, in *Lexikon des Mittelalters*, München und Zürich 1983, Vol. 2, p. 1978f.

imperial guest. In return, Sigismund invited the Queen, together with her ladies-in-waiting and all other noble ladies of her court as well as the most distinguished women of the bourgeoisie, to a banquet of supreme courtly splendour enhanced by a 'concert of harmonious music and admirable voices'.

It is only in his final sentences that the chronicler mentions a political purpose of Sigismund's visit. After the recovery of Charles VI, the imperial guest attended several of the King's council-meetings, where he expressed his grief over the strife between the kings of France and England. Literally the chronicler writes:

> The Emperor added that he would contact the English king, hoping to restore the peace. Thus at the end of April, he graciously said farewell to the King and the officers of his Court and (…) crossed the Channel. (Edition L. Bellaguet, 1844, 747 – my translation).

To sum up, the royal chronicler of Saint-Denis does not make any references to myths or ancient legends. Instead his report reflects the programme of a two-month state visit, which suggests that the designated Emperor perceived Paris and its surroundings as part of a world with which he felt familiar. From the account it also becomes clear that the German sovereign especially admired Paris for its outstanding University, arts, and courtly splendour.

It is interesting to note to what extent the French chronicler reflects the perception of the German King, who, because of his dynastic relations with the French royal family, clearly did not feel alien in the French capital.

The German Knight and Poet Oswald von Wolkenstein (ca. 1376/77-1445) on his Visit to Paris

The situation would have been different for a person such as Oswald von Wolkenstein, who belonged to the low-ranking German nobility without family links to France. But his poetic reflections on the imperial visit to Paris are interesting for yet another reason. Nowadays, the knight Oswald von Wolkenstein is considered the most important German lyric poet between Walther von der Vogelweide and Goethe. He is especially esteemed because he created a new style of autobiographical poetry. With a new language of sensual perception, his autobiographical poems reflect his own experiences, the places he visited, and the people he met, with realistic details of surprising authenticity (S. Hartmann, 2005).

This is also true for the poem Wolkenstein composed about Sigismund's journey to Spain and France (*Es ist ain altgesprochen rat*, edition by Karl Kurt

Klein, 19, 1987). We only have to be aware that this poem is deliberately humorous, ironic, and satirical. It was not composed to praise Sigismund's success but to entertain the imperial entourage. The knightly poet therefore makes a fool of almost everyone, above all of himself.

This may sound like a supreme understatement because by 1415 Oswald von Wolkenstein had reached a first peak in his career as knight and courtly singer.

In February 1415, Sigismund enrolled Wolkenstein in his entourage (A. Schwob, 1999, 223-227). In the spring of the same year, Oswald set out for Portugal, where he took part in the Portuguese conquest of the North African city of Ceuta (E. Koller, 1997, 251-256).

By September, Oswald had joined the imperial party in Perpignan. There he gained the favour of Queen Margaret of Aragon, who awarded him three golden rings for his artistic performance (S. Hartmann, 1997, 133-140). On December 29, 1415, Wolkenstein received King Ferdinand of Aragon's Orden de caballería de la Jarra y Lirios y un Grifo (= the Order of the Knighthood of Our Lady's Vase, Lilies, and Griffin – cf. S. Hartmann, 2001/2002, 322-324); all these outstanding awards were crowned by Queen Isabeau, who honoured his poetry with the gift of a diamond in March 1416 (S. Hartmann, 2001, 60-77).

The whole song about the voyage to Aragon and France (*Es ist ain altgesprochen rat*) consists of 28 strophes, which appears long to read, but is, in fact, extremely entertaining when heard in a musical interpretation.[2] Most of it deals with experiences in Perpignan, Avignon, and Narbonne. This is not really surprising, since Oswald perceived the Spanish customs as exotic and the negotiations as embroiled in cunning ruses. Oswald thus presents a kaleidoscope of episodes seen from behind the stage of the negotiations between the delegates from the Council of Constance and the party of the Antipope before he turns to Queen Margaret and her awards.

XX.

But all that doesn't seem so bad	*Noch ist es ain klainer tadel,*
since the fair Margaret	*seid mir die schöne Margarith*
pierced my ears with a needle,	*stach durch die oren mit der nadel*
which is a custom of her country.	*Nach ires landes sitte.*
That same noble queen	*dieselbe edle künigin,*
fastened two gold rings to them	*zwen guldin ring slos si mir drin*

---

[2] Cf. Musical recording, in: Es fuegt sich. Lieder des Oswald von Wolkenstein. Gesungen von Eberhard Kummer. Preiser Records 1998 with booklet edited by Sieglinde Hartmann and Ulrich Müller, comprising the texts of all songs with an English translation of the song by Alan Robertshaw (quoted).

<table>
<tr><td>

and hung one in my beard;
thus adorned she bade me show off.

</td><td>

und ain in bart verhangen,
Also hiess si mich prangen.

</td></tr>
</table>

XXI.

<table>
<tr><td>

A noble title was bestowed on me:
Viscount of Turkey.
Many took me
for a heathen nobleman.
A splendid Moorish costume, red with
    gold,

was given to me by King Sigismund.
Wearing this, I swaggered about
and sang and danced in heathen style.

</td><td>

Ain edler nam ward mir gelesen:
Wisskunte von Türkei;
Vil manger wont, ich sei gewesen
ain haidnischer frei.
mörisch gewant, von golde rot,

kunig Sigmund mirs köstlich bot,
dorinnen kund ich wol swanzen
und haidnisch singen, tanzen.

</td></tr>
</table>

Without any further introduction, the poet then evokes Paris and the way he experienced the imperial visit, as well as his decoration by the French Queen.

XXII.

<table>
<tr><td>

In Paris thousands of people
thronged the alleys, roads and houses;
men, women and children, a big crowd,

stood for two miles,
all staring at
Sigismund, man of Rome,
and calling me a fool
in my cap and bells.

</td><td>

Zu Paris manig tausent mensch
in heusern, gassen, wegen,
kind, weib und man, ain dick
    gedenns,
stünd wol zwo ganze lege.
die taten alle schauen an
künig Sigmund, römischen man,
und hiess mich ain lappen
in meiner narren kappen.

</td></tr>
</table>

XXIII.

<table>
<tr><td>

The national delegations of all the
    faculties
with their cudgels of gold
paid him greater honour
than an angel as he sat
on a throne in a great hall,
and each school praised him
in a magisterial fashion;
there were students and professors
    without number.

</td><td>

Die nacio von aller schüle

mit iren guldin bengel
erten in auf seinem stüle
noch höher dann ain engel;
und jede schule besunderlich,
die lobt in sicher maisteerlich
in ainem grossen sal,
studenten, maister ane zal.

</td></tr>
</table>

XXIV.

In my old age I learned	*Auf baiden knien so lernt ich gän*
to walk on my knees;	*in meinen alten tagen.*
I didn't dare stand up	*zu fussen torst ich nicht gestän,*
as I approached her:	*wolt ich ir nahen pagen:*
I mean Lady Else of France,	*ich mein frau Elst von Frankereich,*
a most illustrious queen,	*ain künigin gar wirdikleich,*
who with her own hand	*ie mir den bart von handen*
decorated my beard with a diamond.	*verkrönt mit aim diamanden.*

XXV.

In broad lakes you can	*In grossen wassern michel visch*
catch fish by casting nets,	*facht man mit garnen stricken,*
so money was placed on a table before me:	*des ward mir geldes auf ain tisch*
there must have been four-and-a-half big sacks full of it.	*wol fünfthalb grosser secke.*
King Sigismund filled my purse	*künig Sigmund follet mir*
with a heap of silver coins,	*den strich mit manchen plancken zier,*
so that I needed	*was ich an als verzagen*
two others to help me carry it.	*selbdritt neur mocht ertragen.*

XXVI.

Urgent necessity called me away	*Ehafft not mich dar vermüt,*
from there, I had to ride off.	*Von dannen must ich reitten.*
King Sigismund, man of noble blood,	*künig Sigmund, das edel blut,*
would not have me delay.	*schüff pald, ich solt nicht beitten.*
In Paris he gave me his hand in farewell	*von Paris bot er mir die hand*
and sailed over to England —	*und sigelt uber in Engelant,*
to reconcile the kings,	*die künige ze verainen,*
by the way.	*anzu ich das maine.*

(Oswald's text quoted from the edition by Karl Kurt Klein, 19, 1987).

At the end Oswald von Wolkenstein adds a salutary lesson to the whole satire, but it is difficult to decide whether his public took that seriously.[3]

---

[3] For a ground-breaking interpretation see Ulrich Müller: 'Dichtung' und 'Wahrheit' in den Liedern Oswalds von Wolkenstein: Die autobiographischen Lieder von den Reisen. Göppingen 1968 (with translation and commentary). For further secondary literature see Oswald von Wolkenstein. Sämtliche Lieder und Gedichte.

Conclusion

The contrasting depiction of the city of Perpignan with its 'Moorish' atmosphere shows that for German visitors of any social status the city of Paris was not perceived as an exotic place. When Oswald von Wolkenstein makes a fool of himself, it is part of a poetic strategy and does not mean that Paris was a city of fools. On the contrary, the respectful praise of Paris' University and the French Queen reflect the poet's high esteem for the French capital and its royalties. At the same time, the way the German poet characterises himself on his entry into the city does not leave any doubt that Oswald von Wolkenstein tried to continue to play the exotic fool in the French capital as well.

But that was perhaps the poet's way of filling in the gap left by the vanishing of myths, wonders, and legends.

Ins Neuhochdeutsche übersetzt von Wernfried Hofmeister. Göppingen 1989, 380; new translation into modern German, ibid. 61-70.

Bibliography

I. Sources

Chronique du Religieux de Saint-Denys, edited by M. L. Bellaguet, new edition by B. Guenée. 3 vols. Paris 1994.

Jean Froissart: Chroniques. In: Œuvres der Froissart. Kerwyn de Lettenhove (ed.). 25 vols. Bruxelles 1867-1877.

Alexandre Tuetey: Journal d'un bourgeois de Paris (1405-1449). Geneva 1975.

Oswald von Wolkenstein. Liederhandschrift B. Farbmikrofiche-Edition der Handschrift Innsbruck, Universitätsbibliothek, o. Signatur. Introduction by Walter Neuhauser. Munich 1987 (= Codices Illuminati Medii Aevi, 8).

Karl Kurt Klein (ed.): Die Lieder Oswalds von Wolkenstein. 3. edition. Tübingen 1987 (= ATB 55).

Die Lebenszeugnisse Oswalds von Wolkenstein. Edition and Commentary. Anton Schwob et al. (eds.). Vol. 1: 1382-1419. Wien/Köln/Weimar 1999; Vol. 2: 1420-1428. Wien/Köln/Weimar 2001.

II. Secondary Literature

Carla Bozzolo, Hélène Loyau (eds.): La cour amoureuse dite de Charles VI. 2 vols. Paris 1982/92.

Walter Brandmüller: Das Konzil von Konstanz 1414-1418. Band I: Bis zur Abreise Sigismunds nach Narbonne. 2nd edition, F. Schöningh, Paderborn et al. 1999.

Walter Brandmüller: Das Konzil von Konstanz 1414-1418. Band II: Bis zum Konzilsende. F. Schöningh, Paderborn et al. 1997.

Albrecht Classen: The Poems of Oswald von Wolkenstein. An English Translation of the Complete Works (1376/77-1445). New York: Palgrave Macmillan 2008.

Philippe Delorme: Histoire des Reines de France. Isabeau de Bavière. Épouse de Charles VI, mère de Charles VII. Paris 2003.

Jean Favier: Paris au XVe siècle, 1380-1420. Nouvelle histoire de Paris. Paris 1974.

Joerg K. Hoensch: Kaiser Sigismund. Herrscher an der Schwelle zur Neuzeit. Darmstadt 1997.

Sieglinde Hartmann: Sigismunds Ankunft in Perpignan und Oswalds Rolle als *wisskunte von Türkei*. In: *Durch aubenteuer muess man wagen vil*. Festschrift für Anton Schwob zum 60. Geburtstag. Hrsg. von Wernfried Hofmeister und Bernd Steinbauer. Innsbruck 1997 (= Innsbrucker Beiträge zur Kulturwissenschaft; Germanist. Reihe 57). S. 133-140.

Sieglinde Hartmann: Oswald von Wolkenstein und seine Ehrung durch Königin Isabeau von Frankreich (1370-1435). In: Zeitschrift für deutsche Philologie. 120. 2001, S. 60-77.

Sieglinde Hartmann: Ein neues Bildzeugnis Oswalds von Wolkenstein? Die Schutzmantelmadonna von Le Puy-en-Velay und das Marienlied *In Frankreich*. Mit einer kostümgeschichtlichen Untersuchung von Elisabeth Vavra. In: Jahrbuch der Oswald von Wolkenstein-Gesellschaft. Bd. 13. 2001/2002. S. 297-332.

Sieglinde Hartmann: Oswald von Wolkenstein heute: Traditionen und Innovationen in seiner Lyrik, in: Jahrbuch der Oswald von Wolkenstein Gesellschaft, vol. 15. 2005, 349-372.

Sieglinde Hartmann: Pourquoi traduire en français un auteur comme Oswald von Wolkenstein ? In: Translatio litterarum ad penates. Das Mittelalter übersetzen – Traduire le Moyen Âge. Alain Corbellari, André Schnyder et al., eds.. Université de Lausanne 2005, 161-177.

Oswald von Wolkenstein. Sämtliche Lieder und Gedichte. Ins Neuhochdeutsche übersetzt von Wernfried Hofmeister. Göppingen 1989.

La France et les arts en 1400. Les Princes des fleurs de lis. Édition de la Réunion des musées nationaux. Paris 2004.

Ulrich Müller: 'Dichtung' und 'Wahrheit' in den Liedern Oswalds von Wolkenstein: Die autobiographischen Lieder von den Reisen. Göppingen 1968.

Paris 1400. Les arts sous Charles VI. Paris musée du Louvre, 22.03.-12.07. 2004. Exhibition catalogue. Paris 2004.

Erwin Koller: "War der deutsche Ritter etwa Oswald?!" Zur Teilnahme des Wolkensteiners am portugiesischen Überfall auf Ceuta. In: *Durch aubenteuer muess man wagen vil*. Fs. f. Anton Schwob zum 60. Geburtstag. Hrsg. von Wernfried Hofmeister und Bernd Steinbauer. Innsbruck 1997 (= Innsbrucker Beiträge zur Kulturwissenschaft; Germanistische Reihe 57), S. 251-256.

152

M. Pauly und F. Reinert, eds.: "Sigismund von Luxemburg: ein Kaiser in Europa". Tagungsband des internationalen historischen und kunsthistorischen Kongresses in Luxemburg, 8.-10. Juni 2005. Mainz 2006.

Johannes Spicker: Oswald von Wolkenstein. Die Lieder. Berlin 2007 (= Klassiker-Lektüren 10).

I. Takacs: Sigismundus rex et imperator: Kunst und Kultur zur Zeit Sigismunds von Luxemburg 1387–1437 (Sigismund, king and emperor: Art and culture in the age of Sigisumd of Luxembourg 1387–1437). Exhibition catalogue. Mainz, 2006.

Gabrielle Wittkop / Justus Franz Wittkop: Paris. Histoire illustrée. Zurich 1978.

III. CD Recording

Es fuegt sich. Lieder des Oswald von Wolkenstein. Gesungen von Eberhard Kummer. Preiser Records 1998, with booklet, edited by Sieglinde Hartmann and Ulrich Müller.

Prof. Dr. Sieglinde Hartmann
Oswald von Wolkenstein-Gesellschaft
Myliusstr. 25
D – 60323 Frankfurt am Main
E-Mail: wolkenstein.gesellschaft@t-online.de

Foreign Culture in a Foreign Town.
The Nuremberg Poet *Jakob Ayrer* and the Reception of Sixteenth-Century English Comedy Plays in Germany.

The Nuremberg author Jakob Ayrer (ca. 1540-1605) and his contemporary playwrights were inspired by English travelling comedians who first came to perform in the German-speaking parts of Europe in the late sixteenth century. We are not entirely sure, however, exactly how far this influence on Jakob Ayrer went. On the one hand, Ayrer is deeply rooted in local tradition as represented by Hans Sachs; on the other hand, traces of the reception of English drama can clearly be recognised in his dramatic production.[1] The aim of the following study is to show the extension of both, local and English tradition, in Ayrer's works. A look at the Nuremberg Shrovetide plays before Ayrer and at the performances of English comedians in Germany will form the basis for discussing the influence of the English comedians on Ayrer by focussing on the singing plays.

I. Jakob Ayrer and the Nuremberg Shrovetide play tradition

The best-known examples of the early Nuremberg Shrovetide plays, usually set in a single chamber, are the works of Hans Rosenplüt (died after 1460) and Hans Folz (died in 1513). These plays consisted of approximately one hundred to six hundred verses and required between four and ten actors for their performance. Frequently, the column listing the speakers would simply state *der acht, der neunt* and so on. Among the most popular were *Gerichtsspiele*, which frequently focus on sexual desire in the form of *Nachthunger*, and plays about medical doctors, which see defecation as a jolly subject, "als fröhliche Materie" (Bachtin 1995, 36).

---

[1] Bolte (1893, 6) notes: "Ayrer [...] lehnt sich da und dort an das ältere Nürnberger Fastnachtspiel an, wenn er Bauernthorheit vorführt und Volksfiguren wie Eulenspiegel und Claus Narr auftreten lässt," and Baesecke (1935, Repr. 1974, 98) explains: "Die Singspiele und Pickelheringsspiele sind rein stofflich den Fastnachtspielen des Hans Sachs verwandt." On the other hand, Haustein (1993, 581) emphasizes the influence of English troupes: "Ayrer hat von den englischen Komödianten mit ihren Wanderbühnen [...] nicht nur Themen und Vorlagen bezogen, sondern er hat ihnen auch [...] bühnentechnische Neuerungen abgeschaut und ihre gattungsgeschichtlichen Fortentwicklungen übernommen."

154

Departing from this early tradition, the shoemaker and Meistersinger Hans Sachs (1494-1576), whom Richard Wagner called "the greatest master of Nuremberg", arranged his dramas in acts, which cannot be described by the usual laws of poetics, however. The plays stand out because of their precise stage directions and their "konsequente Moralisierung und Didaktisierung" (Bachorski 1992, 10). Frequently Ayrer's dramas are compared to those written by Hans Sachs – usually negatively: Hilsenbeck attests him working "mit großem Eifer", "aber von dem Wesen dramatischer Form hat er keine Ahnung und sein Werk entbehrt der wahrhaft dichterischen Werte"; this reproach for lack of poetic spirit is followed by one for "intrusive" moralization (1950, 117); Ayrer, "mit seinem Hang zum Unwesentlichen, Nebensächlichen und Überflüssigen" (Bock 1971, 285), "übernimmt kritiklos die Sachssche Stoffgier [...]. Während Sachs seinen Stoff bewältigt, erliegt Ayrer der Fabel" (287). Wodick (1912, 40-41) states "unerträgliche Langweiligkeit", and this unbearable dullness leads to Kaulfuß-Disch's (1905, 160) remark that losses through the process of transmission were quite a relief. Opinions on Ayrer's singing plays are in general similarly negative, as Hilsenbeck's reflections (1950, 119) on how the audience could bear their monotony show. Recently, philology has re-read Ayrer's plays with more recognition, and they are considered worthy of a new examination. Haustein (1993, 583) for example denies, with regard to the singing plays, the reproach of monotony with the observation "daß bereits Ayrer durch die Aufteilung der Strophe auf zwei Personen oder die Pointierung der Schlußverse dieser Gefahr entgegenzuwirken versucht hat."

Ayrer, probably born in 1540 or shortly thereafter, the son of a Nuremberg stone carver, was what we today call a career changer: he made a living as an ironmonger in Nuremberg before he moved to Bamberg in 1570, where he took up the business of writing. Ayrer's works comprise a *Tractetlein* (a rhyming chronicle written in couplets) composed in the year 1570, dealing with *Ankunfft, vnd erbauung* of the city of Bamberg. In 1574 he composed a Lutheran translation of the Psalter in verse. In 1593 he had to move back to Nuremberg again and work as an Imperial notary and legal prosecutor. Together with his son, he published the legal work "Belialsprozeß" in Frankfurt in 1597. Ayrer died on March 26th, 1605 in Nuremberg. The posthumous *opus theatricum,* published in 1618, hands down *dreissig ausbündtige schöne Comedien und Tragedien [...] sampt noch andern sechs und dreißig schönen lustigen und kurtzweiligen fassnacht- oder possenspile(n)*, [die man] *persönlich agirn kann.* These extraordinarily fine comedies, tragedies, and lively and amusing shrovetide and singing plays are composed to be performed in the new English manner, *auff die neue Englische manier vnnd art* (foreword, 5-6).[2] As part of a

---

[2] This foreword, composed by *Freunden und Erben* is interesting in that it focusses on

performance manuscript, the stage directions consider everyday dramaturgical needs: for example, that the female protagonist must leave the stage before a fire-blowing dragon comes in order to avoid danger *des schmuckes halben* (K III, 1654); that only an intrepid child should be suitable for the stage (K II, 1248); or that the monetary expenses for musical instruments should be kept within limits (K II, 965).

II. The state of theatre in Germany and English Comedians

In comparison to other European countries, German-speaking theatre was underdeveloped and represented by lay-performers only: "Verglichen mit anderen europäischen Ländern ist Deutschland vor 1600 theatralisches Niemandsland: Fastnachtsspiel, Meistersingerbühne und Schultheater waren Laientheater" (Haekel 2003, 25-30). Thus, the professionalism of the English performers was completely new to the German audience: "So substantial had the impact of these travelling actors been that until the eighteenth century several troupes of purely German actors continued to call themselves 'English Comedians'" (Williams 1990, 29). The continent offered better opportunities for English troupes, which faced fierce competition in their own country. The first English performers came to Dresden in 1586, following an invitation from the elector of Saxony. They arrived from the Danish court, where they had belonged to the entourage of the Earl of Leicester. Among the group was William Kempe, a performer for whom Shakespeare had composed parts as a clown in his plays. In 1592, the group consisting of Robert Browne, John Bradstreet, Thomas Sackville, and Richard Jones arrived at the court of Duke Heinrich Julius of Braunschweig-Wolfenbüttel (1564-1613). At first the performance took place in English (the English plays were not adapted for the German language before 1600). The minor importance of language was balanced by non-linguistic physical expression, such as magnificent gestures and facial expressions, music, dance, improvisation, acrobatics and elements of comedy, because all these could be understood by an audience with little or no knowledge of the English language: "The first time Sackville is mentioned by name he is called a vaulter (der Springer Thomas Sackefiel)" (Schrickx 1986, 190). Therefore, the English

---

the receptors of the work: it might be useful "(in Vermeidung) der Melancholie bei alten betagten Mannspersonen und dem edlen und unedlen Frauenzimmer" (compare Hilsenbeck 1950, 117). The foreword determines the function explicitly in its moral task: *Vnd aber Weyland der Erbar [...] Herr Jacob Ayrer der Elter, Käyserlicher Notarius, [...] hat er damit, seinen Erben vnd Nachkommen zu gleichmessigem abscheu, hergegen aber aller Jugendt ein gutes Exempel zu geben, in seinen lebzeiten* (foreword, 5).

comedians emphasized the visual aspect: "Von den Dramen Shakespeares und der übrigen Dramatiker der Elisabethanischen Epoche bleibt nur das Handlungsgerüst, in dem sich das rein Faktische in blutrünstigen Greueln und manieristischem Pathos entrollt" (Sachs 1957, 127). The number of props was kept very low: "Da das Schauspiel der Englischen Komödianten nicht auf illusionistische Effekte ausgerichtet war, genügten meist wenige Dekorations-gegenstände, um einen Schauplatz zu charakterisieren" (Haekel 2004, 251). Of great importance was "die derbe Aktion mit Prügelszenen, Obszönitäten und drastischer naturalistischer Gestik" (Koch 1974, 2). Such rough action, with scenes of flogging, obscenities, and drastically naturalistic gestures are probably mirrored in the account of the travels of Fynes Moryson; he describes the performance of such a group at the Frankfurt trade-fair:

Germans, not understanding a worde they sayde, both men and wemen, flocked wonderfully to see theire gesture and Action, rather than heare them, speaking English which they understoode not, and pronouncing peeces and Patches of English playes, which my selfe and some English men there present could not heare without great wearysomenes (Moryson, Shakespeare's Europe, cit. Haekel 2003, 76).

Many of the English plays contain no comic parts, but rather comic interludes between acts and scenes. Browne, "so to establish a bond of language between stage and audience, [...] employed a German-speaking clown who summarised the action and entertained by recounting or enacting comically obscene incidents" (Williams 1990, 32). The comic character often became the most important person of the group; a good example is Thomas Sackeville. The effects of his stage character, the clown Johann Bouschet (Jan Bouset), "der ersten komischen Person englischer Prägung auf deutschem Boden" (Haekel 2004, 27) also shows the adoption by Jakob Ayrer and Duke Heinrich Julius of Braunschweig: "Im deutschen Lustspiel wurde die komische Figur unter dem überwältigenden Eindruck der englischen Komiker von jetzt an ein konstituierendes Element" (Catholy 1969, 117). The clown represented the vegetative sphere; it is the figure least dependent on language because he expresses the basic functions of man; this explains the regional differences in naming the clown figure, whose name is usually inspired by nutrition; e.g., Jahn Posset is named after a sweetened spicy drink consisting of cream and beer or wine called *Posset*. Pickelhering and Hans Wurst in German speak for themselves (see Catholy 1969, 119-120).

In German-speaking areas there were no fixed stages, so the travelling performers had to put up with whatever they could get: "any space that would allow an audience to stand, preferably in vertical tiers, around three sides of a temporarily erected stage or in structures resembling a horse corral, which allowed the audience to stand on all sides" (Williams 1990, 29). We can assume that the English performers "were used to making shift with the barest necessity; the best they could hope for was a place used for periodic dramatic entertainments, with the most meagre appointments" (Pascal 1940, 368). The stage would consist of a simple scaffolding with a carpet strung up at the back, two stage exits, and sometimes a balcony. Decoration, word direction, and acoustics show where the story takes place and make clear the non-illusionistic character of the comedy stage: "Die theatralischen Zeichen verweisen nicht mehr auf ein Ereignis, sie sind das Ereignis. [...] Hier kann von einer Dominanz der performativen über der referentiellen Funktion des Theaters gesprochen werden" (Haekel 2004, 256-257). This dominance of the performative over the referential function of the theatre is underlined by the fact that dramas of the travelling English performers do not exist as scripts; they document the actual performance instead of underlying it.[3]

The arrival of a travelling theatre group was a great event for every town. The group of performers had to apply for permission first, which was usually granted for two to three weeks. Performances depended on the paying public, so the whole venture became a business, selling the "Ware Schauspiel" (Haekel 2004, 45; see also Fischer-Lichte 1993, 600), and the performance became a product. The oldest advertising we know about is for a performance in Nuremberg in 1652:

Zu wissen sey jederman daß allhier ankommen eine gantz newe Compagni Comoedianten/ so niemals zuvor hier zu Land gesehen/ mit einem sehr lustigen Pickelhering/ welche täglich agiren werden/ schöne Comoedien/ Tragoedien/ Pastorellen (Schäffereyen) vnd Historien/ vermengt mit lieblichen vnd lustigen interludien/ vnd zwar heut Mittwochs den 21. Aprillis werden sie praesentirn eine sehr lustige Comoedi/ genant

Die Liebes Süssigkeit verändert sich in Todes Bitterkeit
Nach der Comoedi soll praesentirt werden ein schön Ballet/ vnd lächerliches Possensiel. Die Liebhaber solcher Schauspiele sollen sich nach Mittags Glock 2. Einstellen vffm Fechthaus/ allda vmb die bestimbte Zeit praecise soll angefangen werden. (cit. Creizenach 1889, XXV)

---

[3] "Die Dramen der Englischen Komödianten liegen nicht als Spielvorlagen der Wanderbühne vor. [...]. Die Texte waren nicht primär für die Schauspieler der englischen Truppen intendiert, sondern sie wurden für andere Zwecke hergestellt: als Dokumentation einer bestimmten Form von Schauspielpraxis und, in Manuskriptform, als repräsentatives Geschenk. Im Gegensatz zur allgemeinen Vorstellung, sind sie nicht *vor* einer Aufführung, sondern *danach* entstanden" (Haeckel 2004, 100-101); thus they are "der Intention nach die Dokumentation von performativer Praxis, nicht ihr Entwurf" (Haekel 2004, 101).

158

This advertising seems very modern in some respects: The *newe Compagni Comoedianten* claim to be an attraction never seen before in the country. A very funny *Pickelhering* belongs to the team. A daily performance is announced consisting of a various subject matter; what seems most important is the hint at the "lovely and funny" interludes, which points to the function of such funny little plays: to interrupt the main story, perhaps for recreation and relief by laughter. The Nuremberg *Fechthaus* offered a professional building for staging: "For the purposes of dramatic entertainment it may safely be assumed that the stage was built against one side of the building and did not stand detached in the middle of the arena" (Pascal 1940, 369). The advantages are the possibility for players to change clothes, to enter and leave the stage, and to use the gallery, etc. Not the whole side was used for staging, but only one part, "and would jut out into the arena, so that spectators would stand round the stage in the 'pit' as well as look on from the front, side and back galleries" (Pascal 1940, 369). A chronicle of Nuremberg describes the advertising and performance of troupes; equally, there is the complaint about the export of money:

dahin ein groß Zulauffen von Alten vnd Jungen, von Man vnd weibs Personen, auch von Herrn des Raths vnd Doctorn geweßen, den sie mit Zweien trummels vnd 4 trometen in der Statt vmbgangen vnd das volck vfgemohnet vnd ein Jede person solche schone kurtzweillige sachen vnd spiel Zu seien ein halben Patzen geben mueßen, dauon sie, die Comoedianten, ein groß geld vnfgehoben vnd mit ihnen auß dieser Statt gebracht haben. (Trautmann, Englische Komoedianten, 1886, 127, cit. Haekel 2004, 51)

The expression *ein groß Zulauffen* may demonstrate the immense interest of spectators in watching the performance.

III. Ayrer and the English Comedians

The Nuremberg town council ledger attests to the performance of *Ruberto Green* on August 20th, 1593; the actor was probably Robert Browne (according to Robertson 1892, 39-40; and Schricks 1986, 188-189). *Thomas Sachgwde and Companions* were granted permission for four weeks in 1596. For 1596 we know about a performance by the *fürstlichen Hessischen dienern vnd comedianten* [comedians and valets of the prince of Hessia]. In 1597, Sackeville, who had made a considerable impression on Ayrer (see Catholy 1969, 140),[4] returned to

---

[4] Fischer-Lichte (1993, 79) reflects the social possibility that the clown figure offers: "Bezeichnenderweise wurde diese vitale Funktion des Narren zuerst von einem Fastnachtspieldichter erkannt. Jakob Ayrer ergriff nach Jan Bousets erstem Auftreten in Nürnberg sofort die sich hier eröffnende Möglichkeit, mit der Figur seines komischen Dieners Jahn für das von den Stadtvätern Unterdrückte eine

Nuremberg; here, he is already called by his stage name, Jan Bosset. In 1600 the comedians of Hessia performed again, and in 1602 three English groups were given permission to perform, among them Robert Browne's. In 1596 and 1597, Thomas Sackeville was with this ensemble. We know about performances by Robert Browne in 1593, 1596, and 1602 in Nuremberg: "Der Nürnberger Rat, der den einheimischen und auswärtigen deutschen Schauspielertruppen gegenüber stets einen kühlen [...] Ton angeschlagen hatte, kam ihnen [...] fast aufmunternd entgegen" (Herz 1903, 33).

Direct references to English texts occur in only a few of Ayrer's dramas, among them the drama about the *griechischen keyser zu Constaninopel vnnd seiner tochter Pelimperia* (a German version of Thomas Kyd's *Spanish Tragedy*), then the *Comedia vom König in Cypern* (Lewis Machin, The Dumb Knight, in the version of English Comedians), and also in the *Comedia von der schönen Phönicia* (*Much ado about nothing*). Ayrer's *Comedia von der schönen Sidea* contains analogies to Shakespeare's *Tempest*; perhaps an even older English play was the basis for both, as Herz argues (1903, 120).[5] The clown character of the English Jann Posset was adapted by Ayrer sporadically in four of his dramas and completely adopted in fourteen dramas, and it plays a role in ten of the Shrovetide plays (according to Robertson 1892, 49). The singing play *Der engelendische Jann Posset* is partly identical with *Esther* in Ferdinand Menius's collection of English comical plays from 1620 - the drama *Hester and Ahasuerus* was performed by the Lord Chamberlain's group in 1594. Especially with the singing plays the inspiration coming from the English plays can be seen clearly, because Ayrer adopts a total of ten of these singing plays into his repertoire.

IV. The Singing Play *Von dem Engelländischen Jan Posset*

The term *Singspiel* [singing play] refers to any kind of drama, "in welches Musikstücke, vornehmlich Gesänge, sei es in überwiegender, oder auch nur in untergeordneterer Weise eingewebt sind" (Schletterer 1863, 1). The singing plays (jigs) containing music, – songs mainly, to a greater or lesser extent, – were introduced to the German speaking areas by the English. The English singing plays consisted of four or five different types of stanzas, each with an individual tune, similar to well-known songs (see Koch 1974, 3-4). Common

---

Zuflucht und ein Ventil zu schaffen."

[5] The comparison of the *Comedia von der schönen Sidea* (Nr. 28) and Shakespeare's *Tempest* results in the following: "Bei Shakespeare sind die *dramatis personae* in ihren extremen Ausprägungen gut oder schlecht, bei Ayrer hingegen klug oder dumm, bei Shakespeare handeln sie edel, bei Ayrer richtig - richtig und vorteilhaft in ihrem Verhältnis zum Mitmenschen und zu Gott" (Haustein 1993, 578).

topics were the shrewdness of adulterous women, the aggressiveness of women, and love-crazed old people. Not primarily dependent on spoken language, these plays rapidly became favourites with the German audience. The comical figure in the plays providing the interludes between the acts played an important role in achieving this popularity.[6]

According to the Dresden manuscript, Ayrer´s singing plays date back to 1598; ten singing plays have been handed down (see Bolte's list, 1893, 12). Ayrer was not the composer of the music, however; for most of the plays he used the *Engelendischen Rolands Thon*, e.g. for the singing play *Von dreyen bösen Weibern, denen weder Gott noch jhre Männer recht können thun*. This singing play seems to have been Ayrer's first essay in the genre.[7] His singing play *Der engelländische Roland* was first published between 1599 and 1600. The following analysis deals mainly with the singing play *vom engelendische(n) Jann Posset, wie er sich in seinen Dinsten verhalten*, in order to outline parts that can be clearly attributed to the English or the German tradition.

Ayrer changed the story of *Jann Posset* into a Shrovetide play and into a singing play *mit acht Personen, in deß Rolandts Thon*. The last part is of English origin and corresponds, as mentioned before, to the English *Esther*. The shrovetide play, traditionally composed of rhyming couplets, differs formally from the fifty-eight stanzas of the singing play, and this again differs, because it is sung, from the Nuremberg shrovetide plays before Ayrer, when the dramaturgical element of song was used relatively rarely – if we may trust the rare stage directions concerning singing: in the shrovetide play *Of three wicked women* (K 56, 488,3-17) there is a feast with song:

DA TRINKEN SI UND SINGEN UND SEIN FRÖLICH MIT EIN ANDER UND SINGEN DAS GESANK:
[...]
Des tre re ra ro so,/ Des tre re ra ro so.
ODER SI SINGEN DAS:
Lieber wirt, nun trag her wein!"

[THEY DRINK AND SING AND ARE MERRY AND ARE ALL HAPPY TOGETHER:
[...]
Des tre re ra ro so,/ Des tre re ra ro so.
OR THEY SING:
Good landlord, bring the wine]

---

[6] "Es scheint, als sei vor allem der Narr mit seinen Zwischenaktvorstellungen ganz entscheidend an der Einführung des Singspiels beteiligt" (Asper 1980, 26).

[7] Adelbert von Keller, Ayrers Dramen, Fünfter Band, 50-69. Stuttgart 1865 (Bibliothek des Litterarischen Vereins 70), 3051ff. The postscript talks about the first play to sing in the area: *Das ist das erste Spil, / Daß man bey vns hie singen thut.* (3061, 34ff.).

More singing occurs in the Neidhard plays,[8] especially when Neidhard picks violets, as the *Große Neidhartspiel* suggests: *Der Neythart get hyn vnd sucht den veyol frölich mit singen* [Neidhard goes out to pick violets, singing merrily, 37, 655]. Ayrer uses song sporadically in some plays, e.g. in the *possenspil, von einer versoffenen bäurin*: [Diese] *hebt an zu singen, was sie etwa kan, vnd muß darzu juchsen* [she starts to sing whatever she may, and to cheer, 2641, 3-4]. Ayrer does not underlay single scenes with songs to create a comical effect, but rather follows a more elaborate aesthetic concept. Richard Wagner (*Opera and Drama*, 1815, 108) refers to the combination of music and dramatic art in medieval plays as a grown unity with "Spuren eines natürlichen Zusammen-wirkens der Tonkunst mit der Dramatik". By choosing to convey dramatic art in song, Ayrer used this natural interaction of music and dramatic art and was thus able to achieve a great impact on the performative presentation of the play. Performers and audience were linked closely in what Fischer-Lichte (2003, 25, 161) calls a "feedback-Schleife". This feedback loop engenders an energetic effect between players and spectators. In the course of the performance emotions can be created that evoke the danger of "Ansteckung" [infection], which can be strengthened by music. Johann Jakob Bodmer explains the connection between performance and creation of emotions in the spectator in the *Critischen Betrachtungen über die poetischen Gemälde der Dichter* published in 1741.[9] Through the musical presentation, the atmosphere can probably be made denser and the danger of being infected by the play increased.[10]

The content of Ayrer's Shrovetide play and the singing play are – apart from the end of the plays[11] – largely the same: the protagonist Jan is a scoundrel, afraid of work, and making his old parents′ lives a misery with his disrespectful talk. He leaves in anger and takes up work as a farm labourer, at which he is of no real use because he pretends not to understand. Then he gets married, but his wife turns out to be an evil woman who henpecks him till the end. The father of the ill-bred child is aptly called Roland and appears in peasants′-clothes in the singing play. The character of the peasant, who has the ability to construct aesthetics of an upside-down world, appears in Ayrer′s play as well as in the early Nuremberg Shrovetide plays:

---

[8] Margetts, John (Hg.): Neidhartspiele. Graz 1982 (Wiener Neudrucke 7).

[9] J. J. Bodmer, Zürich 1741, Repr. Frankfurt/M. 1971, 290, cit. Fischer-Lichte 2000, 28.

[10] See Fischer-Lichte 2003, 200. See also Gernot Böhme: Atmosphäre. Essays zur neuen Ästhetik. Frankfurt/Main 1995 (edition suhrkamp 1927, NF Bd. 927).

[11] The *moralisatio* differs to a great extent: while the Shrovetide play admonishes all to stay within their own *class*, the singing play focusses on questions of choosing the right spouse and praises an honest wife with whom one can grow old.

162

"Was die Bauernfigur zur beliebtesten Figur des Fastnachtspiels werden läßt bzw. was das Fastnachtspiel als neue Gattung so erfolgreich macht, ist das Prinzip der Verkehrung. Die Möglichkeit, die Normen der offiziellen Welt auf den Kopf zu stellen, ist auch das Gesetz der Fastnacht." (Ragotzky 1983, 97)

In the beginning, literary traditions are evoked; the son wants to leave his family and his class. The subject matter reminds one of Wernhers *Helmbrecht,* and Hans Sachs also employs the motive of the ill-raised son, e.g., in the Shrovetide play *Der ungeraten sun* (KG 6) and in the drama of the *hörnen Sewfriedt* (KG 13),[12] a hero who is not allowed to swear like Jann in Ayrer's singing play:

> *5. Fürwar, alhie so bleib ich nicht.*
> *Ir seit ein grober Baur,*
> *Ir habt ein strenges Angesicht*
> *Vnd secht schellig vnd saur.*
> *So ist die Mutter vngeschaffen,*
> *Zeicht gar zerlampet her,*
> *Runzlet gleich wie die Affen*
> *Vnd brummt als wie ein Beer (2890, 15a-2412)*

> [For sure, I won't stay here.
> You are a rude peasant,
> You have a rough expression
> And you look crazy and sour.
> Mother is really not a beauty,
> She moves about in shabby clothes,
> has wrinkles like the apes
> and she grumbles like a bear.]

The attributes of ugliness in the woman, a part that is traditionally played by a man, helped to enhance the comic effects, as known from the Nuremberg tradition. In the Shrovetide play *Die Bauernheirat,*[13] the bride's father compliments his daughter:

---

[12] In this play, the father complains, *Gott mir ein son bescheret hat, [...] Ist gar unadelicher art,/ Helt zucht und tugendt widerbart,/ Ist frech, verwegen und mutwillig,/ Starck, rüdisch und handelt unbillig;/ Gar keyn höffligkeyt will er lern;/ Es steht all sein gemüt und begern/ Allein zu grobn, bewrischen dingen,/ Zu schlahen, lauffen und zu ringen* (335, 36- 336, 10). Complaints include an ignoble character, disrespect of honour, braggadocio, strength, lack of courtly behavior, a preference for peasant manners, etc.

[13] Keller, Adelbert von (Hg.): Fastnachtspiele aus dem fünfzehnten Jahrhundert. Teil I-III. 1853. Nachlese. Stuttgart 1858 (=Bibliothek des Litterarischen Vereins 28-30; 46).

sie ist von antlutz nit clar,
So hat sie nit vast gelbes har,
So wil ich ir fuß auch nit vast loben,
Aber die pein sind ir gleich unten als oben (71, 9-12)

[her face is not really bright,
her hair is not really blond,
I don't want to praise her feet not much,
but her legs are the same size above and below.]

Having been insulted and beaten, Jann is made to understand the fourth commandment; the father *schlägt den Jannen wol ab* [gives Jannen a good thrashing, 2870, 36] and the stage directions add *zornig*, in rage, a direction familiar from the clerical plays. Both types of plays create a certain habitus that is meant to be made clearly visible for the audience through an act of violence. Especially in the Nuremberg Shrovetide play, e.g. the little Neidhard play, beating is a common way of creating comic effects.

HEBEN STREIT DICIT:
/Hort so heyß ich der heben streit
/ Neithart wer dich wann es ist zeit/
NW SCHLAHEN SIE ANEINANDER
(Margetts 1982, 117, 188-191)

[HEBEN STREIT SAYS:
Listen, my name is quarrel-beginner
Neidhart, defend yourself, because it is time /
NOW THEY BEAT EACH OTHER]

As in the English dramas, Jann takes over the role of a servant, and "the servant was frequently the wit" (Garvin 1923, 1971, 35). When Jan starts to work for the old gentleman Mr. Ermerich, he asks to have his chores written down in a list. This leads to a series of funny misunderstandings in the tradition of Eulenspiegel [Master Tyll Owlglass]: instead of stationery he gets a fire-lighter, instead of a goose quill he brings a cock's plume, then a feather without a sharpened tip, and then he mixes up paper and beer. Because of this he is explicitly referred to as *Eilenspigels Knecht* (2895, 8). The Eulenspiegel tradition is omnipresent in medieval Germany; among others, Hans Sachs wrote Eulenspiegel plays. Words and phrases that are taken literally and that accordingly lead to funny actions also appear in early Shrovetide plays, as the *Fasznachtspil von einem artzt und einem krancken* (K 120), may illustrate:

164

DER ARTZT:
Sag mir her schlecht, wo pistu kranck?
DER KRANCR (sic!):
Secht, mein herr, hie auff diser panck (7, 24-27).

[THE DOCTOR:
Tell me where are you ill?
THE PATIENT:
Look, here on this bench]

In the singing play, when old Mr. Ermerich falls, Jan does not help him up, because it is not among the things written down on his list: *Der alt steht allein auff mit grosser mühe. Jann lacht* [The old man struggles to get up on his own. Jann laughs, 2896, 16a]. This stage direction, including facial expression and gestures, underlines the English comedians' emphasis on emotions and gestures. This stage-direction in all its details and Jann's reaction to the subsequent scolding from the old man – *sicht saur* [looks not amused] – is exceptional for the Nuremberg tradition, as the facial expressions of the performers are hardly ever described. Jann's request for laughter can be read as an attempt to create a "Lachgemeinschaft" [community bound together by laughter]. Ermerich's wish, that Jan should learn courtly manners, reminds one strikingly of Hans Sachs's *hörnen Sewfriedt* (KG 13), who in the beginning refuses to learn any manners at all and then, after his catharsis, decides to work hard on his courtly manners (342,15). Despite all the troubles, Ermerich keeps his servant and tells him to deliver a basket of pears. Jann's ensuing considerations as to whether he should eat the pears himself instead of delivering them (*Jann singt wider sich selber* and *Er besind sich* [Jann sings to himself; Jann reflects a moment]) point to further-developed examples of the German drama, in which soliloquy is frequently used to emphasize the dramatic plot. Jann's eating of the pears is described in detail:

33. O Jann, du wilt es wagen,
Den Piren sprechen zu.
Weils doch niemand thut sagen,
Wie vil ich liffern thu,
Will ich mich drinn ergötzen,
Mir fressen Piren gnug,
Will mich da nidersetzen.
Da hab ich guten fug.
JANN SETZT SICH, FRIST VIL PIRN, GEHT DANN AB,
KUMMT BALT WIDER VND BRINGT NUR EIN PIRN VND SINGT:
34. Ach, sagt mir! hab ich vnrecht thon?
Ein Piren hab ich noch,
Die andern hab ich gfressen schon. (2897, 24-36)

[Oh Jann, you would dare,
to dig in to the pears.
Because no one will tell
How many of them I'll deliver
I will enjoy myself
By eating quite a lot of them
I'll sit down there,
Where I can take a rest
JANN SITS DOWN, DEVOURS MANY PEARS, THEN LEAVES,
COMES BACK SOON WITH ONLY ONE PEAR LEFT AND SINGS:
Oh, tell me, did I do wrong?
There's one pear left
All the others I have already eaten.]

The incorporation of food, described by the term *Fressen* [to gorge] emphasizes the "Körper-Drama," the drama focussing on the body, as Bachtin explains in his considerations on the carnivalization of literature: "das Drama von Geburt, Geschlechtsakt, Tod, Wachsen, Essen, Trinken und Entleerung" (Bachtin 1995, foreword R. Lachmann, 14). Excessive gormandizing is a common topic in the early Nuremberg Shrovetide plays. This applies to both genders alike and is often combined with defecation. A good example is the monologue from the play *die harnaschvasnacht* (K 99, 756, 24-27): *Nu trank ich an einem neuen pier. / Das ich wolt farzen, da beschaiß ich mich gar* [I drank newly brewed beer and when I tried to fart I beshat myself].

Just after Jann decides to get married, his wife seems unhappy with her choice of husband, because *er thut sich stets stellen / Als sey er nicht bey sinn,/Henckt sich an loß Gesellen* [he always pretends to be out of his mind and hangs around good-for-nothing people, 2899, 21-24]. She would like to change her husband: *Vnd will jn jtzt probiren, /Ob er thu fürchten mich* [I will see if he is afraid of me, 2899, 31-32], and she threatens to flog him: *Ich will dirs Maul zerschlagen, / Daß du solst dencken mein* [I will smash your gob, so you know who I am, 2900, 5-6]. Then she makes him carry her shopping:

ELA WÜRFFT DEN KORB FÜR JM NIDER VND SAGT: [14]
Seh, Löll! So faß den Korb balt an!
Vnd was ich kauffe ein,
Daß must du mir heim tragen.
Darumb mir balt nach tritt!

---

14 Thomke (1996, 10833-1084) considers the stage direction "sagt" an error due to the transfer of the shrovetide play into a singing play: not so Catholy (1967, 140), who suggests parts of singing and speaking.

166

[ELA THROWS THE BASKET IN FRONT OF HIM AND SAYS:
See, fool, take the basket immediately!
And all that I buy you are to carry home for me,
So, come after me quickly!]

Here, the props are symbolic of the status of the characters, and the man has to carry the basket. Props and dress in the English comedians' performances depart "von ihrer referenzellen Funktion. Requisit und Kostüm werden theatralischer Selbstzweck" (Haekel 2004, 266-267). The violent woman, as in Ayrer's singing play, calls up misogyny, a way of portraying women common to many medieval texts, employing this topos, as

"entertainment [...] capitalizes on contemporary and perennial prejudices by presenting stereotypic accusations intended for a varied audience, thus catering to, and sometimes ridiculing, misogamous and misogynistic prejudices. The stereotypes of wifely misbehavior in works of general misogamy are funny because their stock characters are predictable, and, through obvious exaggeration, sufficiently removed from reality to appear comic rather than tragic" (Wilson 1990, 6).

Misogyny in portraying women involves sexual greed, a latent potential for violence and combativeness. Women wear *lange klayd vnd kurczn muet* [long dress, but short temper, *ludus de erhardo*].[15] Abuse is so common to women's speech that there are downright tirades of swearing; in the *hubsch[en] vasnachtspil* (K 31, 254, 10-13), the wife accuses her husband of being a *fuller, fresser, saufer und slaucher, speier, zututler, lotter und smaicher, trieger, teuscher, bescheißer und besaicher*, [of gluttony, greed, drunkenness, spitting, laziness, cheating, crapping, and pissing]. The use of swearwords can be attributed to the "mundus perversus" of the shrovetide plays, a "Reservoir, in dem sich die unterschiedlichsten sprachlichen Formen sammelten, die verboten und aus dem offiziellen Sprachgebrauch ausgeschlossen waren" (Bachtin 1995, 67). This pool of various forms of language that were excluded from officially accepted use is not really used in Ayrer's play; the term *löll* for Ayrer's nasty woman is comparatively moderate. The husband is clearly in the weaker position concerning their household situation; his begging for mercy is a motive familiar from early Shrovetide tradition. In the *paurenspil* (K 4, 51, 11-13), the first peasant moans: *Hör auf, liebs weip, es ist zu ser, / Hör auf, schlag nimmer, des pit ich dich* [Stop it, dear woman, it is too much/ oh stop hitting me, I beg you]: "As in most comedy, so in comic and satiric misogamy, the humor derives from a clear incongruity between what is and what should be; between legal and social

---

[15] Bauer, Werner M.: Sterzinger Spiele. Wien 1982 (Wiener Neudrucke. Neuausgaben und Erstdrucke deutscher literarischer Texte 6).

models and situational ethics, on the one hand, and actual human behavior, on the other; between institutionalized male superiority and occurrences of marital *mundus inversus*" (Wilson 1990, 7). Jann sings:

Ey Frau, vmb gnad ich bitt. [...]
ELA SCHLEGT JN AN HALß. JANN SINGT:
Gnad Frau, will ichs doch than; [...]
Ich spinn vnd haspel ab.
Auch will ich gern waschen,
Daß ich eur hult nur hab.

[Ey, lady, I beg your mercy!
ELA BEATS HIM ON THE NECK. JANN SINGS:
Mercy, lady, I'll do it,
I'll take care of yarning and winding up,
And I'll also wash the clothes
Only to receive your favour.]

ELA SINGT:
45. Ich rath dirs auch, du solt es than,
Wilt du vngeschlagen sein.
Hast ein Nasen, die schmecken kan,
So gib dich nur darein!
SIE SCHLEGT JN WIDER. ER WEIND. (2900, 10-34)

[ELA SINGS:
I advise you to do so,
if you don't want to be beaten.
If you have a nose to smell,
then give in!
SHE BEATS HIM AGAIN. HE CRIES.

Keller (2003, 124) asks: "Gibt es komische Männlichkeit, oder eher komische Männlichkeiten?" The question about a comic masculinity or comic masculinities leads to the observation that in this play a kind of masculinity typical of the Shrovetide plays is constructed; the destruction of traditional masculinity is associated with the construction of the comic body. This man's tears can point towards another deconstruction: in contrast to the clerical plays, characters rarely weep or cry in early secular plays (though they often mourn), but there is a weeping clown in Shakespeare's *The Two Gentlemen of Verona*. We might even say that the laughing and weeping clown in Ayrer's work is a novelty leading to later forms of "Lustspiel": in *the Tragedi von der Belägerung Alba vnd den sechs kempfern biss auff den Todt Tullj* [2], stage directions require: *Kompt Jhann, der Bottenlauffer, schreit jammer, greint vnd heilt* (140, 2-3); and later: *Jahnn weind* [Jann, the nuntius, laments, cries, and weeps, 24]. In the *Comedi, Dritter Theil, Von Tarquinio Prisco* [3], we find Jann again in a desolate state: *Jahnn, der Bott, geht ein, stelt sich kläglich vnd schreit: Ach weh, ach weh! was soll ich than?* [Jann, the nuntius, comes in, acts as if in a desolate state and cries: Alas, alas, alas, what shall I do?, 197,33-34]. Later he cries at being laughed at because of his fool's cap (208, 29). In this respect, the *Tragedi, Vierdter Theil, von Servij Tullij* [4] is even more striking. Here, Jodel laughs excessively, and this laughter is even described in its performance and moreover perhaps reflects the spectators' reaction, signalled by the expression "man":

> KOMPT JODEL, [...]
> VNND LACHT, DASS ER ERSCHOTTELT, GEHT MIT SOLCHEM GLÄCHTER ALSO HERUMB.
> DARNACH, WENN MAN SCHIER AUßGELACHT HAT, SAGT ER:
> *Ich het mich schir nacket gelacht* (281, 28-31).

> [JODEL COMES,
> AND LAUGHS SO HARD THAT HE SHAKES, GOES TO AND FRO WITH SUCH LAUGHTER.
> AFTERWARDS, WHEN ONE HAS FINISHED LAUGHING, HE SAYS:
> I nearly laughed my clothes off]

Men doing women's jobs are not uncommon, just as Jann says *I like yarning and winding up and I like to do the washing as well.* In the Nuremberg Shrovetide play *Morischgentanz* (K 14, 123, 21-23) people try to determine who is the biggest fool in love. One of the applicants states the tasks he had to perform for a woman:

> *holz und wasser tragen,*
> *Schussel spulen und wintel waschen;*
> *Ich kert und haizt, hub auf die aschen*

> [carrying wood and water,
> doing the dishes and baby-changing;
> I brushed and heated, took up the ashes]

On the other hand, writers such as Hans Folz and Hans Sachs emphasize the importance of female work. Hans Folz ascertains that: *Die frau ein hauß regiren kan,/ Und ist sorgveltig auff iren man* [a woman can govern the house and take care of her husband, W 9, 71, 231-232]. Hans Sachs in particular emphasizes the task of women in reproduction and the positive aspects of a life with women:

> *Die frawen müssen die welt mehren,*
> *Kinder tragen, ziehen und nehren,*
> *Ds hauß mit allem fleiß verwalten.*
> *Ein fraw die ist ein trost der alten*
> *Und ist der jungn fürsichtigkeit.*

Wer on frawen lebt in der zeit,
Der ist auff erdt lebendig todt.
(*Comedi juditium Salomonis,* Bd. 6, 127, 29-128, 1-4)

[Women are to increase the world,
to carry children, educate, and nourish them,
they look after the house. [...]
He who who always lives without women
is dead on earth while still alive.]

The nominal condition is compared to the actual condition in this play, which shows a threat to the family and the rules of society, as the world that has turned upside down starts to show. Two men turn up who have heard of Jan's fate and want to help him on the basis of a royal decree. Like Hans Knapkäse in *Esther*, Jann is happy "when the king issues the mandate that all wives must obey their husbands. Because of this mandate, and also on the advice of his neighbor, Hans beats his wife [...]. Her submission is not of long duration, however." (Garvin 1923). First, Jann strips Ela of keys and pouch – the regalia of domestic power.

Gib mir schlüssel vnd Beutel her
Vnd trit züchtig nach mir!

Er stürtzt die Arm vnder, sie auch, gehen also rumb.
Die Trabanten lachen vnnd Ela singt:
Ach wie felt mir die sach so schwer!
Der Teufl bscheist mich mit dir. (2902, 14-20)

[Give me the keys and the purse,
And come nicely behind me!
He throws his arms under hers, she does the same,
They turn like this in a circle. The passers-by laugh and Ela sings:
Alas, how hard this is for me!
The devil plays tricks on me with you.]

What research regards as the recipients' "Freude am Bösen" [taking pleasure in evil, thus Röcke 1987] or the "Lust am [...] Übertölpeln von anderen und an der Schadenfreude," (U. Müller 1994, 174) is shifted onto the level of the play by the laughter of the two men. *Itzund bin ich ein Mann* [now I am a man, 2902, 29], Jann rejoices and *gibt jr eins an Kopff* [he hits her on the head]. Here, ideal masculinity is anticipated, because "[d]ie väterliche Autorität inkludiert Gewaltanwendung gegenüber Kindern und Ehefrau, [...] das heißt, Macht und Gewalt sind Teil der Konstruktion vorbildhafter Männlichkeit" (Bennewitz 2000,

17). Beating women is obviously a proper remedy to "cure" women.[16] Jann's newly gained "ideal masculinity" can only be kept up as long as other men are present. After the other two men leave, Ela returns to her old behaviour. This proves the "ambivalente Natur der karnevalistischen Gestalten" by showing that they contain "alle Polaritäten des Wechsels und der Krise" (Bachtin 1990, 53) within them.

SIE GEHEN AB DIE TRABANTEN. ELA WÜRFFT JM DEN KORB WIDER
FÜR VND SINGT: [...]
Seh! da faß den Korb wider an,
Ehe ich dich auch abschmir,
Daß alle Engel lachen,
Weil dein gsellen wegk sein. [...]
SIE REIST JM DEN BEUTEL AUß DEN HÄNDEN, SCHLEGT JHN DAMIT
VBER DIE LEND. JANN SAGT:
Ey, Frau, was wilt du machen?
Ich bitt: schon mein vnd dein! [...]
SIE SCHLEGT JN ZU BODEN, STEHT OBER JM VND SCHMIRT JN WEIDLICH AB. (2903, 6-27)

[The passers-by leave. Ela throws the basket again in front of his feet and sings:
Look there! Take the basket again,
To avoid me thrashing you in a way
that makes all the angels laugh,
Because your comrades have left
SHE VIOLENTLY PULLS THE PURSE OUT OF HIS HANDS, BEATS HIM WITH IT AROUND THE
LOINS. JANN SINGS:
Hey, lady, what are you up to?
I beg you: pity me and you!
SHE KNOCKS HIM TO THE GROUND, STANDS OVER HIM AND THRASHES HIM WELL.]

In Ayrer's singing play *Ein schoens neus singets Spil von dem Knoerren Cuentzlein mit vier Personen*,[17] the female protagonist Lampa uses similar phrases as she becomes violent against her husband: *Lampa nimbt den Pesen, schmirt den Mann weidlich ab* [Lampa grabs the broom and gives her husband a

---

[16] Violence against women seems to be apt to create comic feelings for Ayrer. In the *possenspil, von einer versoffenen bäurin, wie sie vmv jhren kram vnnd kleider betrogen vnd jhrem mann fast nacket heimgeschickt ward* [K 40], there is beating, too; here a cleric recommends that the husband of a drunken wife cures her by beating her: *Als dann so will ich dich wol lehrn/Ein salb, damit du sie solst reibn,/ Die versuffen weiß zu vertreibn.* (2650,19-21)

[17] Wuttke calls it "Die Erziehung des bösen Weibes" [the taming of the wicked wife], using the tune: *Venus, du und dein Kind*; Böhme, Altdeutsches Liederbuch Nr. 219; see Wuttke [6]1998, 288.

good thrashing, 305, 378a]. In this play, a neighbour watches the action and comments on it with obvious pleasure in evil – which is not to be recommended.

64. O dran, dran, Nickel dran!
Die Frau die schlegt den Mann,
Sie schlegt in mit eim Pesen [...]
Ich mein, sie hab ein wesen.

LAMPA LAUFFT AUFF IHN UND SINGT
Du Schelm, halt das Maul zu,
Ehe ich dir auch so thu!
SI SCHLEGT IHN AUCH UNND GEHT AB (305, 379-384a)

[64. O go on, do it, do it!
The woman's beating the man,
she's beating him with a broom
She sure has an attitude.

LAMPA RUNS TO HIM AND SINGS
You scoundrel, shut your gob,
Before I do the same to you!
SHE BEATS HIM, TOO, AND LEAVES]

In the Nuremberg shrovetide plays, women often show a great disposition for violence; a single woman can easily stand up to two men, as in the *paurenspil* [play of the peasants]: *Die zwen pauren fliehen unter die pank und das weib erwischt den ersten pauren* [The two peasants run and hide under a bench but the woman gets hold of the first one (K 4, 50, 24-26)]. The victim of female violence calls for help:

DAS WEIB WIRFT DEN PAUREN NIDER UND SCHLECHT IN, DAS ER ALSO SCHREIT:
O helft, lieben freunt, pox leichnams willen! (K 4, 50, 9-10)

[THE WOMAN THROWS THE PEASANT TO THE FLOOR AND BEATS HIM UNTIL HE CRIES OUT:
Oh help me, dear friends, for the love of Christ!]

Jann calls for help in a similar way: *O helfft, jr Nachbaurn mein!* [Oh help me, my companions!, 2903, 6-29], and the two men come to his rescue. Overall this does not create the desired effect and leads to an unusual moral at the end of the play:

ELA SAGT: [...]
Ich will euch bed abschmirn.
SIE SPRINGT AUFF, SCHLEGT DIE BEDE TRABANTEN VND JREN MANN AB VND SINGT:

57. Also habt jr vernommen,
Wie ein jungs Weib allein
In der schlacht vberkommen
Hat sich gewehret fein
Dreyer Männer wol einmal.
Darumb, wer freyen will
Vnd nicht kommen in pein vnd qual,
Der lerne auß dem spil (2904, 15-27)

[ELA SAYS:
I will thrash all of you.
SHE JUMPS TO HER FEET, BEATS THE TWO PASSERS-BY AND HER HUSBAND AND SINGS:
So all of you have heard
Of a young lady who has won the battle
on her own and defended herself well
against three men with a single blow.
So he who plans to marry
And doesn't want trouble and pains,
should let himself be advised by the play.]

After she has been thrashed by her husband, the female protagonist Lampa in Ayrer′s singing play *Die Erziehung des bösen Weibes*, announces: *von Hertzen gern/ Will ich [...] Dir folgen in alln sachen* [with all my heart I will follow you wherever you go, 308, 451-453] and so is successfully tamed, as are most wicked women in medieval literature (e.g., Sibote, Frauenzucht). In Ayrer's singing play about Jann Posset, the wife remains in power, Ayrer uses this unusual outcome to thwart the expectations of the listener. While Jann Posset's Shrovetide play advises staying in one's social class, the singing play focusses on the process of picking a woman in the first place, praising the one one can grow old with. This is a positive aspect emphasized by Hans Sachs as well: *Bey der man kan wol alten* [A woman is a comfort for the old, *Comedi juditium Salomonis*, 128, 1].

V. Conclusion

It becomes evident that Ayrer's singing play *Von dem Engelländischen Jann Posset* uses all the means of comic effects to be found in the Nuremberg Shrovetide play before him. Traditional elements are abuse, thrashings, excessive eating, swearing, intentional misinterpretations of words, and means of contrast. The well-known Shrovetide play topic of the "wicked wife" is taken up in the form of a nagging, violent woman. On the other hand, the singing play differs from the Nuremberg tradition. Ayrer seems, in many respects, more moderate. The aggressive obscenities of the early Nuremberg plays and the comedy based overwhelmingly on excrement no longer appear with this intensity. The same can already be said for Hans Sachs.

In any case, Ayrer's often-ignored plays contain much worldly wisdom. As the aggrieved squire in the singing play states - and this could be regarded as lifestyle motto - "I am laughing, so I should be swearing" (2875, 16). Ayrer's text stands out particularly through its appearance as a singing play. The tradition of the English comedians also appears in the content of Ayrer's singing play, which shows similarities to the drama *Esther*. Influences can also be seen in elaborate facial expressions and gestures, which do not appear in such detail in Hans Sachs' dramas. The crying and weeping in Ayrer's plays remind one strikingly of Shakespeare's weeping clown. New impulses, such as the comic character Jann Posset, gain popularity. Through the use of pre-existing subjects (Boccaccio) and the influence of English comedians, the secular play is led to an endpoint. From 1650 on, English Comedians and German travelling performers unite.

174

Bibliography

I. Sources

Keller, Adelbert von (Hg.): Fastnachtspiele aus dem fünfzehnten Jahrhundert. Teil I-III. Stuttgart 1853. Nachlese. 1858 (Bibliothek des Litterarischen Vereins 28-30; 46).

Keller, Adelbert von (Hg.): Ayrers Dramen, Vierter Band, Stuttgart 1865; Fünfter Band. Stuttgart 1865 (Bibliothek des Litterarischen Vereins 69; 80).

Keller, Adelbert von/ Goetze, Edmund (Hg.): Hans Sachs. Sämtliche Werke, 26 Bde. Tübingen 1870-1908, Bd. 6 (1872), Repr. Hildesheim 1964; Bd. 14 (1882), Repr. Hildesheim 1964.

Margetts, John (Hg.): Neidhartspiele. Graz 1982 (Wiener Neudrucke 7).

Thomke, Hellmut (Hg.): Deutsche Spiele und Dramen des 15. und 16. Jahrhunderts. Frankfurt/M. 1996 (Bibliothek der Frühen Neuzeit, Erste Abteilung: Literatur im Zeitalter des Humanismus und der Reformation Bd. 2; Bibliothek deutscher Klassiker 136).

Wuttke, Dieter (Hg.): Fastnachtsspiele des 15. und 16. Jahrhunderts. Unter Mitarbeit von Walter Wuttke ausgewählt und herausgegeben. Stuttgart 1973, 2. Aufl. 1978 (RUB 9415).

II. Secondary Literature

Asper, Helmut G.: Hanswurst. Studien zum Lustigmacher auf der Berufs-schauspielerbühne in Deutschland im 17. und 18. Jahrhundert. Emsdetten 1980.

Bachtin, Michail: Literatur und Karneval. Zur Romantheorie und Lachkultur. Frankfurt/M. 1990.

Bachtin, Michail: Rabelais und seine Welt. Volkskultur als Gegenkultur. Frankfurt/M. 1995 (suhrkamp tb 1187).

Baesecke, Anna: Das Schauspiel der englischen Komödianten in Deutschland. Seine dramatische Form und seine Entwicklung. Halle/S. 1935 (Studien zur englischen Philologie 87). Repr. Walluf/Wiesb. 1974.

Bauer, Werner M.: Theater. In: Deutsche Literatur. Eine Sozialgeschichte. Hg. von Horst Albert Glaser. Bd. 2: Von der Handschrift zum Buchdruck. Reinbek b. Hamburg 1991, S. 81-115.

Bennewitz, Ingrid: Frühe Versuche über alleinerziehende Mütter, abwesende Väter und inzestuöse Familienstrukturen. Zur Konstruktion von Familie und Geschlecht in der deutsche Literatur des Mittelalters. Jahrbuch für internationale Germanistik 32, 2000, H. 1, 8-18.

Bock, Hans B.: Jakob Ayrer. In: Fränkische Klassiker. Eine Literaturgeschichte in Einzeldarstellungen hg. von Wolfgang Buhl. Nürnberger Presse 1971.

Bolte, Johannes: Die Singspiele der englischen Komödianten und ihrer Nachfolger in Deutschland, Holland und Skandinavien. Hamburg, Leipzig 1893. (Theatergeschichtl. Forschungen 7).

Catholy, Eckehard: Das Fastnachtspiel des Spätmittelalters. Gestalt und Funktion. Tübingen 1961 (=Hermaea. N.F. 8) (Tübinger Habilitationsschrift 1958).

Catholy, Eckehard: Das deutsche Lustspiel. Vom Mittelalter bis zum Ende der Barockzeit. Stuttgart u.a. 1969.

Creizenach, Wilhelm: Die Schauspiele der englischen Komödianten, Berlin, Stuttgart 1889.

Fietz, Lothar/Fichte, Joerg O./ Ludwig, Hans-Werner: Semiotik, Rhetorik und Soziologie des Lachens. Vergleichende Studien zum Funktionswandel des Lachens vom Mittelalter zur Gegenwart. Tübingen 1996.

Fischer-Lichte, Erika: Kurze Geschichte des deutschen Theaters. Tübingen, Basel 1993 (UTB 1667).

Fischer-Lichte, Erika: Entgrenzung des Körpers. Über das Verhältnis von Wirkungsästhetik und Körpertheorie. In: Fischer-Lichte, Erika/Fleig, Anne (Hg.): Körper-Inszenierungen. Präsenz und kultureller Wandel. Tübingen 2000, 19-34.

Fischer-Lichte, Erika: Ästhetik des Performativen. Frankfurt/M. 2003 (edition suhrkamp 2373).

Flemming, Willi: Das Schauspiel der Wanderbühne. Stuttgart 2. Aufl. 1931. Repr. Darmstadt 1965 (Deutsche Literatur in Entwicklungsreihen. Reihe Barock. Barockdrama 3).

Garvin, Wilhelma: The development of the comic figure in the German drama from the Reformation to the Thirty Years' War. Philadelphia 1923. New York 1971.

Le Goff, Jacques: Das Lachen im Mittelalter. Mit einem Nachwort von Rolf M. Schneider. Aus dem Französischen von Jochen Grube. Stuttgart ^{2}2004.

Haekel, Ralf: Wanderbühne. In: Erdmann, Eva (Hg.): Der komische Körper. Szenen – Figuren – Formen. Bielefeld 2003, 25-30.

Haekel, Ralf: Die Englischen Komödianten in Deutschland. Eine Einführung in die Ursprünge des Berufsschauspieltums. Heidelberg 2004.

Haustein, Jens: Jacob Ayrer. In: Füssel, Stephan (Hg.): Deutsche Dichter der frühen Neuzeit. 1450-1600. Ihr Leben und Werk. Berlin 1993, 575-588.

Herz, E.: Englische Schauspieler und englisches Schauspiel zur Zeit Shakespeares in Deutschland. Hamburg, Leipzig 1903. (Theatergeschichtliche Forschungen XVIII). Repr. Nendeln 1977.

Hilsenbeck, Fritz, in: Nürnberger Gestalten aus neun Jahrhunderten. Ein Heimatbuch zur 900-Jahrfeier der ersten urkundlichen Erwähnung Nürnbergs. Hg. von Stadtrat zu Nürnberg. Nürnberg. Nürnberg 1950, 116-120.

Hinck, Walter: Das deutsche Lustspiel des 17. und 18. Jahrhunderts und die italienische Komödie. Commedia dell'arte und théâtre italien. Stuttgart 1965 (Germanistische Abhandlungen 8).

Kaulfuß-Disch, Carl Hermann: Die Inszenierung des deutschen Dramas an der Wende des sechzehnten und siebzehnten Jahrhunderts. Ein Beitrag zur älteren Bühnengeschichte. Leipzig 1905 (Probefahrten 7).

Keller, Johannes: Dekonstruierte Männlichkeit: von scheintoten und begrabenen Ehemännern. In: Aventiuren des Geschlechts. Modelle von Männlichkeit in der Literatur des 13. Jahrhunderts. Hg. von Martin Baisch, Hendrikje Haufe, Michael Mecklenburg, Matthias Meyer und Andrea Sieber. Göttingen 2003 (Aventiuren 1), 123-148.

Koch, Hans-Albrecht: Das deutsche Singspiel. Stuttgart 1974 (M 133).

Kowalski, Edward/Wegner, Michael: Michail M. Bachtin: Untersuchungen zur Poetik und Theorie des Romans. Berlin, Weimar 1986.

McDonald, William C.: Mythos Eulenspiegel – Sieg eines zwitterhaften Listreichen. In: Müller, Ulrich/ Wunderlich, Werner (Hg.): Verführer, Schurken, Magier, Mittelaltermythen 3. Konstanz 2001, 227-241.

Merkel, Johannes: Form und Funktion der Komik im Nürnberger Fastnachtspiel. Freiburg/Br. 1971 (Studien zur deutschen Sprache und Literatur 1).

Müller, Ulrich: Zur Lachkultur in der deutschen Literatur des Mittelalters: Neidhart und Neidhart Fuchs. In: Laughter down the Centuries Bd. I. Hg. von Siegfried Jäkel und Asko Timonen. Turku 1994 (Annales Universitatis Turkuensis Sarja – Ser. B OSA – 208 Humaniora), 161-181.

Pascal, R.: The Stage of the "Englische Komödianten" – Three Problems. In: Modern Language Review 35 (1940), 367-376.

Probst, Hans: Jakob Ayrer und Bamberg. Neues über sein Leben und seine Werke. In: Historischer Verein Bamberg 85 (1937), 5-27.

Ragotzky, Hedda: Der Bauer in der Narrenrolle. Zur Funktion ,verkehrter Welt' im frühen Nürnberger Fastnachtspiel. In: Typus und Individualität im Mittelalter. Hg. von Horst Wenzel. München 1983 (Forschungen zur Geschichte der älteren Deutschen Literatur 4), S. 77-101.

Ragotzky, Hedda: Pulschaft und Nachthunger. Zur Funktion von Liebe und Ehe im frühen Nürnberger Fastnachtspiel. In: Ordnung und Lust. Bilder von Liebe, Ehe und Sexualität in Spätmittelalter und Früher Neuzeit. Hg. von Hans-Jürgen Bachorski. Trier 1991, S. 427-446).

Robertson, J.G.: Zur Kritik Jakob Ayrers mit besonderer Rücksicht auf sein Verhältnis zu Hans Sachs und den Englischen Komödianten. Diss. Leipzig 1892.

Röcke, Werner: Die Freude am Bösen. Studien zu einer Poetik des deutschen Schwankromans im Spätmittelalter. München 1987 (Forschungen zur Geschichte der älteren deutschen Literatur 6).

Röcke, Werner: Schälke – Schelme – Narren. Literaturgeschichte des 'Eigensinns' und populäre Kultur in der frühen Neuzeit. In: Schelme und Narren in den Literaturen des Mittelalters. XXVII. Jahrestagung des Arbeitskreises Deutsche Literatur des Mittelalters (Greifswald), Eulen-spiegelstadt Mölln, 24.-27.09.1992 (Wodan Bd. 31. Serie 3, Bd. 16), S. 131-149.

Röcke, Werner/Velten, Hans R.: Lachgemeinschaften. Kulturelle Inszenierungen und soziale Wirkungen von Gelächter im Mittelalter und in der Frühen Neuzeit. Berlin, New York 2005 /Trends in Medieval Philology 4).

Sachs, Hans Günter: Die deutschen Fastnachtspiele von den Anfängen bis zu Jacob Ayrer. Diss. (Masch.Schr.) Tübingen 1957.

Schletterer, Hans Michael: Das deutsche Singspiel von seinen ersten Anfängen bis auf die neueste Zeit. Augsburg 1863. Hildesheim, New York 1975.

Schrickx, Willem: Foreign Envoys and Travelling Players in the Age of Shakespeare and Jonson. Wetteren 1986.

Simon, Eckehard: Die Anfänge des weltlichen deutschen Schauspiels. 1370-1530. Untersuchungen und Dokumentation. Tübingen 2003 (MTU 124).

Styan, J.L.: The English Stage. A History of Drama and Performance. Cambridge 1996.

Trautmann, Karl: Englische Komoedianten in Nürnberg bis zum Schlusse des Dreissigjährigen Krieges (1539-1648). In: Archiv für Litteraturgeschichte 14 (1886), 127.

Wade, Mara R.: The German Baroque Pastoral 'Singspiel'. Bern u.a. 1990 (Berner Beiträge zur Barockgermanistik 7).

Wagner, Richard: Oper und Drama (1851). In: Material zum Theater. Texte zur Ästhetik, Dramaturgie und Aufführungspraxis der deutschen Oper im 18. und 19. Jahrhundert. Zusammenstellung und Einleitung von Stephan Stomper (Beiträge zur Theorie und Praxis des sozialistischen Theaters 65, Reihe Musiktheater 14).

Williams, Simon: Shakespeare on the German Stage. Vol. I: 1586-1914. Cambridge 1990.

Wilson, Katharina M. / Makowski, Elizabeth M.: Wykked Wyves ant the Woes of Marriage. Misogamous Literature from Juvenal to Chaucer. New York 1990.

Wodick, Wilibald: Jakob Ayrers Dramen in ihrem Verhältnis zur einheimischen Literatur und zum Schauspiel der englischen Komödianten. Halle/S. 1912.

Dr. Andrea Grafetstätter
Otto-Friedrich-Universität Bamberg
Lehrstuhl für deutsche Philologie des Mittelalters
An der Universität 5
D - 96047 Bamberg
E-Mail: andrea.grafetstaetter@uni-bamberg.de

Heavenly Jerusalem as a *locus amoenus* in Medieval and Early Modern Polish Literature

The purpose of this study is to investigate how the category of a locus amoenus, Latin for 'pleasant place', could be applied to old Polish literary descriptions of the Heavenly Jerusalem, and to analyse its medieval and early modern Polish literary devices.

To limit the brief summary of characterisation of this traditional topos to the most obvious bibliographic entries,[1] it is enough to say that Curtius's term generally refers to a beautiful and shaded segment of nature with grass, trees, and water, an idealised place of safety or comfort. The literary usage of this type of setting, in Western literature at least, goes back to Homer's Odyssey, Virgil's Eclogues, Theocritus's Idylls, or Ovid's Metamorphoses. And although over the centuries of its use, the topos of locus amoenus developed certain limited expressive capabilities that tended to circumscribe its usage, it constituted a powerful, universally recognizable rhetorical commonplace of rest, relaxation, and retirement.[2]

The Christianization of the *locus amoenus*

Obviously, by the late Middle Ages, the topos had undergone many changes and been transformed significantly, reflecting the theological conditions that informed a predominately Christian culture.[3] Departing from the concept of the locus amoenus as a rural haven from political strife, Christian writers began to depict the "pleasant place" as a sensuous and permanent heavenly paradise.[4] Given the biblical account of Eden (Genesis and Song of Songs) and New Jerusalem (Revelation) as places of perfection and harmony, such a trans-

---

[1] Ernst Robert Curtius, *European Literature and the Latin Middle Ages*, trans. Willard R. Trask, (New York, 1953), pp. 183 ff.

[2] Cf. David Evett, "Paradice's Only Map": The "Topos" of the "Locus Amoenus" and the Structure of Marvell's "Upon Appleton House", PMLA, Vol. 85, No. 3. (May, 1970), p. 506.

[3] Paul Piehler argues that Patristic and medieval Biblical exegesis provided a certain key to the interpretation of the appearance of the *locus amoenus* in any medieval literary text. (*The Visionary Landscape. A Study in Medieval Allegory*, London 1971, p. 79).

[4] Thomas P. Harrison, ed., *The Pastoral Elegy: An Anthology*, New York, 1968, pp. 6-8.

formation – the gradual shift in the locus amoenus away from natural consolation and towards divine providence – was almost inevitable[5].

Although Curtius limits the meaning of the locus amoenus to the presentation of landscape alone, without collapsing the distinction between city and locus amoenus, I would like to consider them together. Since the time of the Fathers, the interpretation of the Heavenly Jerusalem of Apocalypse has been a complex one. Following the fourfold medieval hermeneutic, "the City" was understood on different levels in different ways.[6] Here the anagogical understanding of earthly Jerusalem stands for the upper City, New Jerusalem.

The words from St. John's Book of Revelation set out a vision of heaven that has captivated the Christian imagination.[7] The description in the final book of the Bible was the main source for the literary images of Heavenly Jerusalem in old Polish literature, too.[8] The literary evidence sourced here is generally early modern, but the ideas that these texts embody can also be found in late Polish medieval texts.

---

[5] Cf. Robert Bernard Hass, *The Mutable Locus Amoenus and Consolation in Tennyson's In Memoriam, Studies in English Literature, 1500-1900*, Vol. 38, No. 4, (Autumn, 1998), p. 672.

[6] Robert D. Russell, *"A Similitude of Paradise": The city as Image of the City*, in: *The Iconography of Heaven*, edited by Clifford Davidson, Kalamazoo, Michigan 1994, p. 149.

[7] E.g. *Vision of Thurkil of Essex, Vision of the knight Tondal, Le Pèlerinage de la vie humaine* by Guillaume de Deguilleville, Hildegard's reflections in her canticle "O Jerusalem, aurea civitas", Middle-English *Pearl*, or seventeenth-century *Heavenly City* by John Bunyan, to name but a few. Cf. Alister E. McGrath, *A Brief History of Heaven*, Blackwell Publishers, Oxford 2003, chapter: *The City: The New Jerusalem.*

[8] "I saw the Holy City, the New Jerusalem, coming down out of heaven from God... The angel who talked with me had a measuring rod of gold to measure the city, its gates and its walls. The city was laid out like a square, as long as it was wide." *(Rev.21:2, 15-17)* "The wall was made of jasper, and the city of pure gold, as pure as glass. The foundations of the city walls were decorated with *every kind of precious stone*. The first foundation was jasper, the second sapphire, the third chalcedony, the fourth emerald, the fifth sardonyx, the sixth carnelian, the seventh chrysolite, the eighth beryl, the ninth topaz, the tenth chrysoprase, the eleventh jacinth, and the twelfth amethyst" *(Rev.21:17-20)* "It had a great, high wall with twelve gates, and with twelve angels at the gates. (...) The wall of the city had twelve foundations, and on them were the names of the twelve apostles of the Lamb... The twelve gates were twelve pearls, each gate made of a single pearl." *(Rev.21:12-14, 21)*

The locus amoenus has traditionally had both sensual and moral functions.[9] Although its sensual function was favoured in classical literature, its moral function has found vivid expression in the literature of Christian authors. Besides, it is obvious that raising the question of the urban form in a literary text requires one to address the more general question of "space" in literature and its textual function.

The Heavenly Jerusalem as a *locus amoenus* in Old Polish Literature

As Alister McGrath argues, to speak of heaven is to affirm that the human longing to see God will one day be fulfilled – we shall finally be able to gaze upon the face of what Christianity affirms to be the most wondrous sight anyone can hope to behold.[10] But as Saint Paul wrote: "Eye hath not seen, nor ear heard, neither have entered into the heart of man, the things which God hath prepared for them that love him." (1 Cor 2:9). This statement was a source of a conventional "inexpressibility topos," meaning that heaven is impossible both for the reader to imagine and for the author to describe. Many Polish authors used this ritual confession of the powerlessness of words in describing the Heavenly Jerusalem. Peregryn of Opole in his *Sermones de tempore et de sanctis*, ca. 1297-1304, wrote:

> "Si totum coelum est pergamenum et totum mare incaustum et omnes stelle essent magistri Parisienses et omnia stramina penne, certe hi omnes magistri nec manibus nec lingua possent describere minimum gaudium, quod sancti habent de vultu Dei."[11]

Fortunat Łosiewski in his 'Memoriał kaznodziejski' wrote in a similar vein:

---

[9] Michael Squires, *Adam Bede and the Locus Amoenus*, Studies in English Literature, 1500-1900, vol. 13, no 4, Autumn 1973. p. 671.

[10] McGrath, *op. cit.*

[11] Peregrini de Opole, *Sermones de tempore et de sanctis*. E codicibus manu scriptis primum edidit Richardus Tatarzyński. Institutum Thomisticum PP. Dominicanorum Varsaviensium 1997. [If the entire sky were a parchment, all the sea was ink, all stars were the scholars of Paris and all straws were the feather pens, certainly neither those scholars, nor their hands nor tongues would be able to describe the smallest joy that saints have from at the sight of God's face. - transl. JK]. On the figure of the parchment and the ink to express the infinite joys of Heaven see: I. Linn, If All the Sky were Parchment, PMLA, vol. 53, no. 4, (Dec. 1934).

182

"Stańże mi tu Archimedesie, a zmierz obszerność i wspaniałość tego Świętego Miasta górnej Jerozolimy, wysokość wież, moc murów i wałów, dróg i ulic, bram wspaniałość opisz. Aleć darmo myśleć o tym, lepiej nic nie mówić o tym Mieście, aniżeli ni to, ni owo, albo bardzo mało. (s. 82). [...] Bo takie są radości i ukontentowania w niebie, że choćby cały firmament stał się tak wielkim karteluszem jak sam jest, całe morza atramentem, zioła i frukta piórmi, wszystkie stworzenia manualistami stały się, wyrazić ich i opisać nie potrafią."[12]

Translation:

Oh, Archimedes, measure the size and the glory of the eternal city, upper Jerusalem, the height of its towers, strength of its walls and embankments, describe the beauty of its streets and gates. But it is pointless to even think or try to picture the glory of this city. You'd better remain silent, because you cannot say much about this magnificent city. Because there is so much joy, so much happiness in heaven, that, if the entire sky became an enormous parchment, all the seas became ink, all herbs and fruits were pens, and all creatures were writers, they could not express and describe the beauty of this city.

Likewise, Klemens Bolesławiusz in 'The Terrifying Echo of the Last Trumpet' (*Przeraźliwe echo trąby ostatecznej*) wrote:

"Które jakie jest nikt tego nie powie,
Aż sam zobaczy, dopiero się dowie;
Co Bóg zgotował swoim przeznaczonym,
Błogosławionym.
Nie tylko język, lecz i myśl ustaje,
Gdy chce uważyć, jakie tam są kraje
W nagrodę hojną świętym pozwolone,
Wiecznie poszczone."[13]

Translation:
One cannot express the beauty of the city until one sees it; then one will get to know what God has prepared for the blessed. Not only tongues but also thoughts cannot manage to comprehend the holy reward for those who will be allowed to enter the eternal city.

---

[12] F. Łosiewski, *Memoriał kaznodziejski rzeczy ostatecznych wszystkim prawowiernym podany...*, Kraków 1736, p. 79. All translations by Jacek Kowzan, unless stated otherwise.

[13] K. Bolesławiusz, *Przeraźliwe echo trąby ostatecznej...*, Poznań 1670, p. 117-118.

Nevertheless, the "inexpressibility topos" was a convention that did not restrain writers from describing the beauty of the New Jerusalem awaiting the blessed. So despite admitting to the impossibility of the task, writers do discuss the celestial city at length:

"Cóż jest, co widzę? Miasto nieprzejrzane,
Królestwo wielkie, nieoszacowane,
Splendor i chwałę niewypowiedziane,
Nieopisane.
[...] Domy, pałace, o jako z drogiego
Kamienia wszytkie dyjamentowego,
W tych zaś pokoje topazyjuszowe.
Hijacyntowe.
Rynki, o jakie wielkie i szerokie!
Gmachy prześliczne, przestronne, wysokie,
Wszytkie kamieńmi sadzone jasnemi,
O , jak drogiemi!
Ulice jako kształtnie rozłożone,
Wszytkie, o jako pięknie rozmierzone!
W każdej pełno jest przedziwnej piękności
I wesołości."[14]

Translation:
What can I see? A city unseen, the great and priceless realm. Its splendour and glory are ineffable, inexpressible. Houses and palaces made of precious stones and diamonds, with rooms of topaz and hyacinth. What large and wide squares! What beautiful, spacious and high buildings, all set with such bright and precious stones! All streets, how well-laid out, well-proportioned, well-designed! In each of them there is much exceptional beauty and joy.

A comparison with the Apocalypse shows that Bolesławiusz's approach to "the City" is a poetic paraphrase of the relevant biblical verses.

Although St. John's description of New Jerusalem is quite detailed, writers reworked traditional material in order to provide the contemporary reader with a more accurate image of the celestial City; this would meet their requirements, their knowledge, imagination, and experience.

As the state of happiness has to be fulfilled in a particular place, it requires consideration of the space where the souls will be rewarded. In comparison with the heavenly city, the earthly metropolises are belittled and considered as low ones. Franciszek Łosiewski wrote:

---

[14] Bolesławiusz, *op. cit.*, p. 122.

184

"I wspaniała Kartagina, i sławna Babilonia, i rozwlekła Niniwa, złota Aleksandria, uczone Ateny, Konstantynopol poważny, Ekbatana szlachetna, Rzym Pani świata i insze by najsławniejsze Miasta do Niebieskiej przyrównane Jerozolimy, są to dla żółwiów skorupy, dla jaskółek gniazda, dla mrówek usypana jama, albo ślimaczy domek. [...] Gdyby cała ziemia odmieniła się w złoto, woda w balsam wezbrała, góry zmieniły się w drogie kamienie, to wszystko jak nic w komparacyi do Miasta Niebieskiego." [15]

Translation:
And magnificent Carthage, and famous Babylon, and great Nineveh, golden Alexandria, learned Athens, distinguished Constantinople, noble Ecbatana, and Rome, The Maiden of The World, and other famous ancient cities, when compared with the Heavenly Jerusalem are like turtles' shells, swallows' nests, ants' mounds or snails' shells. And if the whole earth were transformed into gold, water into balm, and mountains into precious stones, they would be nothing in comparison to The Heavenly City.

Likewise, Antoni Węgrzynowicz's hyperbolic comparison to the cities of that time, for example Florence, Rome, Venice, Bologna and Cracow, emphasizes the qualities of Heavenly Jerusalem. According to Rosalind Field, the city as a locus has as much poetic power as the garden, and the heavenly Jerusalem is for medieval and early modern poets the pattern of which the Troy, Rome, Camelot, or Sarras of legend are but shadows. [16]

Moreover, Franciszek Łosiewski argues, the Paradisiacal City is perfect because it was designed and built according to the outstanding and unlimited eternal wisdom of God, the Architect: "Cives coelestis patriae/ Regi regum concinite/ Qui supernus est artifex/ Civitatis uranicae / In cuius aedificio/ Talis exstat fundatio." (Anselm of Laon, Commentary on the Apocalypse). [17]

Therefore, the eternal and perfect city cannot be destroyed by a flood, or by an earthquake or a battle. This is a glorious city because of its name, the City of God, the Lord's residence and the dwelling of His saints.

Generally, in Old Polish texts the authors attempted to represent "the place" as having a physical reality. It was because human language finds itself pressed

---

[15] Łosiewski, *op. cit.*, p. 82, 85.

[16] Rosalind Field, *The Heavenly Jerusalem in "Pearl"*, Modern Language Review, 81:1 (1986: Jan.), p. 16.

[17] *Patrologia Latina* 162, 1580. [The residents of the heavenly home, sing for the King of the Kings who is celestial artist, in his edifice such a foundation rises. Transl. – JK] See also S. Kobielus, *Człowiek i ogród rajski w kulturze religijnej średniowiecza*, Warszawa 1997, s. 13.

to its limits when trying to depict and describe the divine.[18] Therefore words and images are borrowed from everyday life and put to new uses in an attempt to capture and preserve precious insights into the nature of God and His City.[19] Since the purpose of the literary descriptions of the Heavenly City were more symbolic than topographical, the standard image settled on was compact and self-contained, yet still possessed all the necessary parts of the city.[20] These urban descriptions always included beautiful walls constructed of bright and precious materials.[21] They gleamed like gold and were often full of gems and precious stones, as, for example in Sermones super Gloria in excelsis by Stanislaw of Skarbimierz:

"Ibi civitas regis magni, in qua per omnes vicos canitur Alleluia, cuius plateae stratae sunt auro mundo, portae ex margaritis purissimis, in qua presto sunt omnes lapides pretiosi et strepitus in ea nunquam audiuntur. Illius civitatis mansiones saphiris fundatae, laterculis aureis coopertae, qua nullus ingredietur immundus, nullusque potest habitare pollutus". [22]

Translation:
There is the community of the great King, in which through all the places they sing Hallelujah, the streets are made of gold, the gates made of the purest pearls, where all stones are precious and no one can hear the bustle. The houses of this city are made of sapphire and covered with golden roof tiles, where no one who is impure can enter and dwell.

The Heavenly Jerusalem as *urbs quadrata* and the Garden of Eden

Despite the general conventionality, there were still some specific relationships between these literary images of the Heavenly Jerusalem and the description found in the Apocalypse of St. John.

Early Polish writers who took St. John's description as their starting point invariably selected certain aspects of the scripture, rather than dealing with the impossibility of rendering in literary form what the biblical text described. More

---

[18] Cf. McGrath, *op. cit.*

[19] *Ibidem.*

[20] Robert D. Russell, *op. cit.*, p. 146.

[21] Cf. Hana Šedinová, *The Precious Stones of Heavenly Jerusalem in the Medieval Book Illustration and Their Comparison with the Wall Incrustation in St. Wenceslas Chapel*, Artibus et Historiae, Vol. 21, No. 41. (2000), pp. 31-47.

[22] Stanisław ze Skarbimierza, *Sermones super "Gloria in excelsis"*, ed. R. M. Zawadzki, Warszawa 1978.

186

frequently, however, the writers picked, chose, and assembled a vision of the Heavenly City made up of what they took to be its most important elements. Not surprisingly, these features tended to be the things that were important to the representation of earthly cities: walls, towers, and first and foremost the quadratic form.[23]

Geometric perfection of "the City" is based on the square. The square represents fairness, balance, and firmness. Something that is "squared" is something that is stable, a foundation for building upon. Therefore *urbs quadrata* is a symbolic image of the incommutable eternal happiness that constitutes the locus amoenus.

However, as Stanisław of Skarbimierz argues in his sermon, not everyone has access to the Heavenly Jerusalem. Only the blessed who present "the passport of virtue" at the door of St Peter may enter the Heavenly City. This reveals the moral (tropological) and anagogical vein of the description.

In Old-Polish texts, Heavenly Jerusalem as a locus amoenus is mainly constituted by walls. It refers to a space that is supposed to be relatively closed, limited, and sheltered. The dialectic "open-closed" is clearly relevant here, because the enclosure not only protects from danger and excludes the Profane, but also provides safety and the inclusion of the Sacred. Therefore the Heavenly City is distinctly conceived not only as a place of amenity, but also as a refuge from the processes of time and mortality; it offers respite from danger, sickness, pain, and sadness. An interesting characteristic of such descriptions is the use of positive statements alongside negative ones to describe an absence of unpleasant things – there is no snow or hail, no heat or cold, no thirst or hunger.[24]

Alternatively, the Polish medieval preacher Stanisław of Skarbimierz contended in his Sermones super "Gloria in excelsis" that heaven was a place of safety, abundance, divinity, and delight and offered the highest form of consolation:

"Ibi mansio est secura, continens totum, quod delectat. Est enim civitas Regis illius plena divitiis, superabundans deliciis." [25]
Translation:
There's a safe shelter that includes everything that delights. For there is a community of the King, full of riches and superabundant delicacies.

---

[23] King James Bible: "And the city lieth foursquare, and the length is as large as the breadth: and he measured the city with the reed, twelve thousand furlongs. The length and the breadth and the height of it are equal." (Rev 16, 21).

[24] This element of description can be found among others in the texts of Piotr Skarga (*Kazania o siedmiu skaramentach… Kazania przygodne…Kazania o czterech rzeczach ostatecznych…*, Krakow 1600, p. 438), Bolesławiusz and others.

[25] *Ibidem.*

Various literary devices were often combined in order to depict the abstract concept of the City of God as a pleasant place. In Old Polish literature locus amoenus was sometimes presented in contrast with locus horridus; and portae coeli were often described in contrast to the mouth of hell.

The utmost joys and happiness discussed here are from the secondary blessings. As I am concerned mostly with the spatial forms of a locus amoenus I leave aside the essential blessing, including the community of saints and the beatific vision.

Most of the Old Polish texts I have examined point out, following the Bible, that "the City" has no need for the sun to shine upon it, for the glory of God is its light. What is certain is the importance Polish authors assigned to "harmony and light" in the attainment of the beatific vision. As geometric harmony and light have such importance in the medieval system of thought, it is logical that they played a significant role in literature as well.[26]

In late medieval and early modern Polish literature the "garden of paradise" became an integral element of the landscape of Heaven, which "increasingly assumed a pastoral quality". Thus Heaven is often described in a mixed way that includes both urban and natural elements. The latter are derived from descriptions of places such as the Garden of Eden and are characterised by an abundance of flowers, fruits, meadows, fountains, and trees to please the senses of sight and smell, and sweet singing birds to please the ears. These are, as we can see, the essential features of the locus amoenus according to Curtius's classic definition. Therefore the paradisiacal City and the Garden of Eden seem to be two interrelated and apparently inseparable elements of the landscape of Heaven in early Polish literature. Such a mixing of a garden (i.e. nature) and a city (i.e. culture) was meant in the Middle Ages as a sacralization of the former.

These elements form a draft picture of the Heavenly Jerusalem as a *locus amoenus*. Elaborations, especially in poems, were confined mainly to these depictions, and the total image was generally concluded by a recapitulation, in condensed form, of the depictions elaborated upon in the locus amoenus itself.[27]

---

[26] Cf. Laurence Hul Stookey, *The Gothic Cathedral as the Heavenly Jerusalem: Liturgical and Theological Sources*, Gesta, Vol. 8, No. 1. (1969), p. 39.

[27] Frederic C. Tubach, *The Locus Amoenus in the "Tristan" of Gottfried von Straszburg*, Neophilologus, vol. 43, no 1, December 1959, p. 38.

Conclusion

After taking early Polish descriptions of Heaven as a frame of reference, we may conclude, as these examples have demonstrated, that varied devices were combined in order to imagine the abstract concept of the City of God and thus to manifest the Vision of John. However, Heavenly Jerusalem as a *locus amoenus* did not remain unattainable because it existed only in language. According to both authors and readers it was not merely a poetic convention, but a real place that can be reached through a virtuous life and death.

The Heavenly Jerusalem was an archetype of the good to come, which would be revealed on the redemption of the world. However, fulfilment can only come through achievement; therefore the celestial city pictured in literature was an uplifting symbol for the mind. Described in sermons, ascetic treatises, and devotional poems, it acted as a rhetorical argument to make listeners and readers eager to reach heaven in the afterlife. The topos therefore became a vessel for essentially religious feelings. Mostly the authors concerned themselves with keeping their poetic imagination flexible enough so that they could explore new possibilities of hope and redemption in the paradisiacal city.[28]

---

[28] Hass, *op. cit.*, p. 673. Cf. J. Kowzan, *Quattuor hominum novissima. Dzieje serii tematycznej czterech rzeczy ostatecznych w literaturze staropolskiej*, Siedlce 2003, *passim*.

Bibliography

I. Sources

Bolesławiusz, K.: Przeraźliwe echo trąby ostatecznej (...), Poznań 1670.

Łosiewski, F.: Memoriał kaznodziejski rzeczy ostatecznych wszystkim prawowiernym podany (...), Kraków 1736.

Opole, Pegrini de: Sermones de tempore et de sanctis. E codicibus manu scriptis primum edidit R. Tatarzyński. Institutum Thomisticum PP. Dominicanorum Varsaviensium 1997.

Rader, M., J. Niesius: Quattuor hominis ultima, Poznań 1648.

Skarbimierza, Stanisław ze: Sermones super "Gloria in excelsis", ed. R. M. Zawadzki, Warszawa 1978.

Skarga, P.: Kazania o siedmiu sakramentach (...), Kraków 1600.

Węgrzynowicz, A.: Kazań niedzielnych księga trzecia, Częstochowa 1713.

II. Secondary Literature

Bernet, Claus: Das Himmlische Jerusalem im Mittelalter. In: MEDIAEVISTIK. Vol. 20, 2007, 9-35.

Curtius, E. R.: European Literature and the Latin Middle Ages, transl. by W. R. Trask, London 1953.

Evett, D.: "Paradise's Only Map": The "Topos" of the "Locus Amoenus" and the Structure of Marvell's "Upon Appleton House", PMLA, Vol. 85, No. 3. (May, 1970).

Hass, R. B.: The Mutable Locus Amoenus and Consolation in Tennyson's In Memoriam, Studies in English Literature, 1500-1900, Vol. 38, No. 4, (Autumn, 1998).

Field, R.: The Heavenly Jerusalem in "Pearl", Modern Language Review, 81:1 (1986: Jan.)

Kobielus, S.: Człowiek i ogród rajski w kulturze religijnej średniowiecza, Warszawa 1997.

Kowzan, J.: Quattuor hominum novissima. Dzieje serii tematycznej czterech rzeczy ostatecznych w literaturze staropolskiej, Siedlce 2003.

190

McGrath, A. E.: A Brief History of Heaven, Blackwell Publishers, Oxford 2003.

Michałowska, T.: Średniowiecze, Warszawa 1995.

Russell, R. D.: "A Similitude of Paradise": The city as Image of the City, in: The Iconography of Heaven, edited by Clifford Davidson, Kalamazoo, Michigan 1994.

Schulze-Belli, P.: "From the Garden of Eden to the *locus amoenus* of Medieval Visionaries" in: Hartmann, Sieglinde (ed.): Fauna and Flora in the Middle Ages. Studies of the Medieval Environment and its Impact on the Human Mind. Papers Delivered at the International Medieval Congress, Leeds, in 2000, 2001 and 2002, Frankfurt am Main, Berlin, Bern, Bruxelles, New York, Oxford, Wien, 2007, 209-224.

Šedinová, H.: The Precious Stones of Heavenly Jerusalem in the Medieval Book. Illustration and Their Comparison with the Wall Incrustation in St. Wenceslas Chapel, Artibus et Historiae, Vol. 21, No. 41. (2000)

Sokolski, J.: Pielgrzymi do piekła i raju. T.I. Świat średniowiecznych łacińskich wizji eschatologicznych, Wrocław 1995.

Sokolski, J.: Staropolskie zaświaty. Obraz piekła, czyśćca i nieba w renesansowej i barokowej literaturze polskiej wobec tradycji średniowiecznej, Wrocław 1994.

Sokolski, J.: Architektura zaświatów w średniowiecznych łacińskich wizjach eschatologicznych, [in Wyobraźnia średniowieczna, pod red. T. Michałowskiej, Warszawa 1996.

Squires, M.: Adam Bede and the Locus Amoenus, Studies in English Literature, 1500-1900, vol. 13, no 4, Autumn 1973.

Stookey, L. Hul: The Gothic Cathedral as the Heavenly Jerusalem: Liturgical and Theological Sources, Gesta, Vol. 8, No. 1. (1969)

Tubach, F. C.: The Locus Amoenus in the "Tristan" of Gottfried von Straszburg, Neophilologus, vol. 43, no 1, December 1959.

Dr Jacek Kowzan
Akademia Podlaska w Siedlcach
Podlaska Academy in Siedlce
ul. Orzechowa 53/10
50-540 Wroclaw
Poland
jacek.kowzan@wp.pl

Beihefte zur Mediaevistik
Monographien, Editionen, Sammelbände

Herausgegeben von Peter Dinzelbacher

Band 1 Albrecht Classen: Verzweiflung und Hoffnung. Die Suche nach der kommunikativen Gemeinschaft in der deutschen Literatur des Mittelalters. 2002.

Band 2 Ralph Frenken: Kindheit und Mystik im Mittelalter. 2002.

Band 3 Werner Steinwarder: Romanische Kunst als politische Propaganda im Erzbistum Lund während der Waldemarzeit. Studien, besonders zum Bild der Heiligen Drei Könige. 2003.

Band 4 Werner Heinz: Musik in der Architektur. Von der Antike zum Mittelalter. 2005.

Band 5 Gudrun Wittek (Hrsg.): *concordia magna*. Der Magdeburger Stadtfrieden vom 21. Januar 1497. 2006.

Band 6 *Mai und Beaflor*. Herausgegeben, übersetzt, kommentiert und mit einer Einleitung von Albrecht Classen. 2006.

Band 7 Olaf Wagener/Heiko Laß (Hrsg.): *...wurfen hin in steine/grôze und niht kleine...* Belagerungen und Belagerungsanlagen im Mittelalter. 2006.

Band 8 Sieglinde Hartmann (ed.): Fauna and Flora in the Middle Ages. Studies of the Medieval Environment and its Impact on the Human Mind. Papers Delivered at the International Medieval Congress, Leeds, in 2000, 2001 and 2002. 2007.

Band 9 Christa Agnes Tuczay: Ekstase im Kontext. Mittelalterliche und neuere Diskurse einer Entgrenzungserfahrung. 2009.

Band 10 Olaf Wagener (Hrsg.): Der umkämpfte Ort – von der Antike zum Mittelalter. 2009.

Band 11 Olaf Wagener / Heiko Laß / Thomas Kühtreiber / Peter Dinzelbacher (Hrsg.): Die imaginäre Burg. 2009.

Band 12 Fabian Rijkers: Arbeit – ein Weg zum Heil? Vorstellungen und Bewertungen körperlicher Arbeit in der spätantiken und frühmittelalterlichen lateinischen Exegese der Schöpfungsgeschichte. 2009.

Band 13 Elisabeth Mégier: Christliche Weltgeschichte im 12. Jahrhundert: Themen, Variationen und Kontraste. Untersuchungen zu Hugo von Fleury, Ordericus Vitalis und Otto von Freising. 2010.

Band 14 Andrea Grafetstätter / Sieglinde Hartmann / James Ogier (eds.): Islands and Cities in Medieval Myth, Literature, and History. Papers Delivered at the International Medieval Congress, University of Leeds, in 2005, 2006, and 2007. 2011.

www.peterlang.de

Peter Lang · Internationaler Verlag der Wissenschaften

Rudolf Suntrup / Jan R. Veenstra (Hrsg.)

Himmel auf Erden – Heaven on Earth

Frankfurt am Main, Berlin, Bern, Bruxelles, New York, Oxford, Wien, 2009.
XX, 201 S., 12 Abb.
Kultureller Wandel vom Mittelalter zur Frühen Neuzeit / Medieval to Early Modern
Culture. Herausgegeben von Volker Honemann und Martin Gosman. Bd. 12
ISBN 978-3-631-56420-2 · br. € 39,80*

Für den mittelalterlichen Menschen waren das Bewusstsein und der Lebensalltag
vom Glauben an das konkret vorgestellte Jenseits in einer Selbstverständlichkeit
und Intensität geprägt, die für den modernen Menschen kaum noch nachvoll-
ziehbar ist. Wenngleich die Bildsprache vom ‚Himmel' und die Rede vom ‚Himmel
auf Erden' heute gerade im außerreligiösen Kontext – als Buchtitel, im Schlager,
in Redensarten und im Werbetext – verbreitet ist, wird in diesem thematisch
geschlossenen Tagungsband in Einzelstudien der Frage nachgegangen, wie im
Spätmittelalter und im Übergang zur Frühen Neuzeit die Idealvorstellung vom
‚Himmel' auf bestimmte Formen individueller Lebensführung, gesellschaftlicher
Organisation und künstlerischer Gestaltung einwirkt. Konkret fassbar wird dies
etwa in Bereichen von Politik und Gesellschaft (Herrscher, Staat, Schulwesen,
Theokratie), in religiöser Praxis (zweckbestimmte Armenfürsorge, Wallfahrt) und in
bestimmten Kunstformen (Meistergesang, geistliches Lied, allegorische Dichtung).
Der Band enthält sechs deutschsprachige und drei englische Beiträge.

For medieval man heaven was a concrete reality. Belief in the afterlife was self-
evident and intense in a way that is difficult to imagine for modern man who
knows heaven sooner from booktitles, songs, figures of speech or advertisements
than from every-day experience. The contributions to this volume of proceedings,
however, deal with the question how in the late-medieval and early-modern period
the idealized image of heaven influenced life, society and art. The various essays
deal with the impact of this idealism on politics and society (ruler, state, education,
theocracy), on religious practice (poor relief, pilgrimage), and on different art forms
(Meistergesang, religious song, and allegorical poetry). The volume contains six
German and three English contributions.

Frankfurt am Main · Berlin · Bern · Bruxelles · New York · Oxford · Wien
Auslieferung: Verlag Peter Lang AG
Moosstr. 1, CH-2542 Pieterlen
Telefax 00 41 (0) 32 / 376 17 27

*inklusive der in Deutschland gültigen Mehrwertsteuer
Preisänderungen vorbehalten
Homepage http://www.peterlang.de